SPINOFF

Innovative Partnerships Program

2005

Developed by
Publications and Graphics Department
NASA Center for AeroSpace Information (CASI)

Front Cover Image:

Artist Lynette Cook's concept of a Jupiter-like planet orbiting the star 55 Cancri (about 41 light-years from Earth). In the background, at left, is the star, and in the foreground is a hypothetical icy moon.

National Aeronautics and
Space Administration

For sale by the U.S. Government Printing Office
Superintendent of Documents, Mail Stop: SSOP, Washington, DC 20402-9328
ISBN 0-16-075266-3

Table of Contents

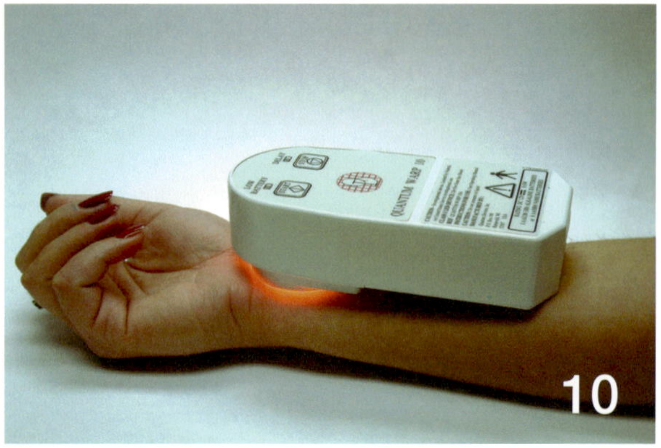

66

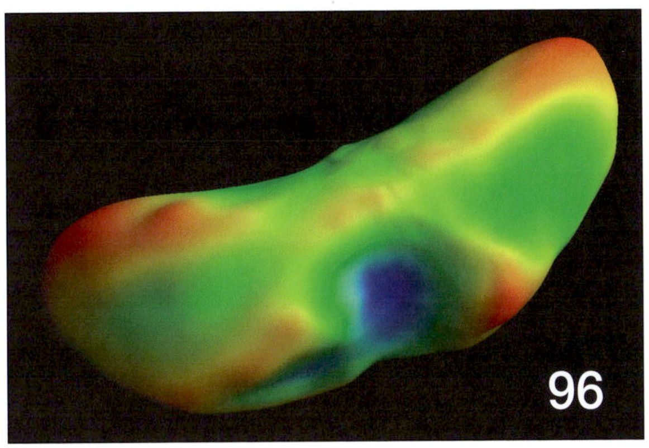
96

114

Foreword

On January 14, 2004, President George W. Bush announced the Vision for Space Exploration, giving the National Aeronautics and Space Administration (NASA) a new and historic focus and clear objectives.

The fundamental goal of the Vision is "...to advance U.S. scientific, security, and economic interests through a robust space exploration program." In issuing this directive, the President committed the Nation to return human explorers to the Moon by the end of the next decade, and to prepare for the exploration of Mars that will follow. NASA is now working hard to develop a new generation of spacecraft and space launch vehicles that will enable the achievement of these goals within the modest expenditure of tax revenues—on average, $55 per year for every American citizen—that our Nation invests in space exploration and research.

As we continue to explore the universe, I am confident that NASA's pioneering exploration activities will keep fueling American creativity, innovation, and technology development. Indeed, throughout the Agency's history, technologies developed to advance our exploration missions have boosted economic progress and benefited millions of people here on Earth.

Spinoff 2005 highlights NASA's work, consistent with our Agency's charter, to "research, develop, verify, and transfer advanced aeronautics, space, and related technologies." Among the beneficial NASA-derived technologies featured in *Spinoff* 2005 now utilized in the commercial and public sector are:

- a bacterial spore-detection unit designed to sterilize Mars-bound spacecraft that can also recognize anthrax and other harmful spore-forming bacteria.
- a remote command and control system NASA uses to run experiments on the International Space Station that allows people to use a cell phone, personal digital assistant (PDA), or Internet connection to activate their kitchen appliances and begin cooking dinner before they get home.
- space suit technology used in the production of lighter-than-air vehicles, such as blimps and dirigibles, during pharmaceutical manufacturing and the production of gas masks for military and civilian use.
- a prototype of the Mars Exploration Rover that is being used in Afghanistan and Iraq to help U.S. troops clear caves and bunkers, search buildings, cross live antipersonnel mine fields, and deal with the dangers posed by improvised explosive devices.
- lightning-detection devices used on NASA's launch pads at the Kennedy Space Center in Florida that are now being used to pinpoint lightning strikes at airports.
- a powerful lubricant designed for use in turbomachinery that is now being widely used in industry.
- a filter designed for use on satellites being used to clean the air breathed by racecar drivers.

These innovations demonstrate that a vigorous space exploration program has and will continue to provide the American public with an impressive technological return on investment. Although technological spinoffs are ancillary benefits of NASA's exploration activities, and not the chief reason for doing what we do, they are tangible and benefit the country.

Michael D. Griffin
Administrator
National Aeronautics and Space Administration

Introduction

As we begin our journey to establish a sustained human presence in the solar system, we continue to be excited by the technical challenges NASA faces today. NASA Administrator, Michael Griffin, put it best when he said, "The President's directive for the Vision for Space Exploration gave all of us who are privileged to work in this business a challenge bold enough to last a lifetime."

Since 1976, NASA has produced *Spinoff* magazine. It was created to highlight the Agency's most significant research and development activities as well as the successful transfer of NASA technology. It shows not only the cutting-edge research being done by the Nation's top technologists, but also the practical benefits that come back down to Earth in the form of tangible products that make our lives better. The stories in the Partnership Benefits section focus on NASA technologies being used by the public today.

How, then, do these spinoffs originate? Each spinoff starts with a NASA mission, which drives the development of new technologies in order to make the mission a success. NASA partners with industry, academia, and other Federal entities to jointly develop technologies, and mature and test the technologies using shared laboratories, test beds, and facilities. Through these partnerships, all sides can leverage one another's ideas and investments that lead to new capabilities and benefits. One benefit of such partnerships is that the technologies can often be applied outside of a mission in various markets.

Many talented people at NASA and its partnering entities are responsible for making these partnerships successful. Those working on behalf of NASA include NASA researchers, engineers, and contractors who create technologies for NASA, plus the Innovative Partnerships Program staff, licensing agents, and legal counsel who make the partnerships work. NASA's partners are technologists, new business developers, and the business talent who conceive opportunities and create new products. These experts bring with them a wealth of information from diverse backgrounds to help transfer the technologies from the laboratory to the marketplace. They are the experts with the technical and business acumen needed to bring these industry partnerships to fruition.

When the combination of NASA and industry technology culminates in new uses for technologies, these applications evolve into the jewels that become spinoffs. For example, a Space Shuttle pump component found application in the development of a heart pump; a charge coupled device developed for the Hubble Space Telescope found application in noninvasive breast biopsy procedures; and alternative energy sources pioneered by NASA are leading the way in advances in the use of renewable resources.

Developing technologies of interest to NASA and the commercial marketplace is an exciting joint venture. The results form a bond among NASA, industry, and the American public. The outcome of this innovative, entrepreneurial process creates value that is a concrete dividend for America's investment in the Space Program. *Spinoff* captures many of these examples.

It is with great pride that we present to you *Spinoff* 2005. Each year, we feature NASA's most significant technological achievements, made possible not only through the research and development, but also through the many talented individuals behind the scenes. We hope you enjoy reading this year's edition and learning how space research and exploration have had a positive impact on the economy and on our lives.

Merle McKenzie
Acting Director
Innovative Partnerships Program

PARTNERSHIP BENEFITS

NASA seeks to create industry partnerships to develop technology that both applies to NASA mission needs and contributes to commercial competitiveness in global markets. As part of NASA's statutory charter, the Agency facilitates the transfer and commercialization of NASA-sponsored research and technology. These efforts not only support NASA, they enhance the quality of life here on Earth.

Lighting the Way for Quicker, Safer Healing

Originating Technology/ NASA Contribution

Who's to say that a little light can't go a long way? Tiny light-emitting diode (LED) chips used to grow plants in space are lighting the way for cancer treatment, wound healing, and chronic pain alleviation on Earth.

In 1993, Quantum Devices, Inc. (QDI), of Barneveld, Wisconsin, began developing the HEALS (High Emissivity Aluminiferous Light-emitting Substrate) technology to provide high-intensity, solid-state LED lighting systems for NASA Space Shuttle plant growth experiments. The company evolved out of cooperative efforts with the Wisconsin Center for Space Automation and Robotics (WCSAR) at the University of Wisconsin-Madison—a NASA center for the Commercial Development of Space. Ronald W. Ignatius, QDI's president and chairman, represented one of WCSAR's industrial partners at the time. WCSAR was conducting research on light sources for promoting food growth within closed environments where humans would be present for a long duration, such as the Space Shuttle and the International Space Station.

With the support of WCSAR, Ignatius experimented with LEDs, which provide high-energy efficiency and virtually no heat, despite releasing waves of light 10 times brighter than the Sun. Ignatius admits that some scientists involved in the project were skeptical at first, thinking that the idea of using LEDs to promote plant growth was far-fetched. The experiments, however, demonstrated that red LED wavelengths could boost the energy metabolism of cells to advance plant growth and photosynthesis. This finding prompted Ignatius to develop a line of LED products that emit the exact wavelength of light that plants use in photosynthesis.

"Our company gives credit to Dr. Ray Bula, the director of WCSAR, for having the foresight to go against the prevailing dogma of the time and design the first plant experiment using monochromatic light to grow lettuce plants," Ignatius proclaims.

In 1989, Ignatius formed QDI to bring the salt grain-sized LEDs to market, and in October 1995, the light sources made their Space Shuttle flight debut on the second U.S. Microgravity Laboratory Spacelab mission (STS-73, Columbia).

Partnership

When NASA determined that red LEDs could grow plants in space, Marshall Space Flight Center awarded QDI several **Small Business Innovation Research (SBIR)** contracts to investigate the effectiveness of the broad-spectrum diodes in medical applications. The contracts, issued from 1995 to 1998, focused on increasing energy inside human cells. It was NASA's hope that the LEDs would not only yield medical benefits on Earth, but that they would help to stem the loss of bone and muscle mass in astronauts, which occurs during long periods of weightlessness. (In space, the lack of gravity keeps human cells from growing naturally.) Furthermore, since wounds are slow to heal in a microgravity environment, LED therapy could accelerate healing and keep what would be termed as minor wounds on Earth from becoming mission-catastrophic in space.

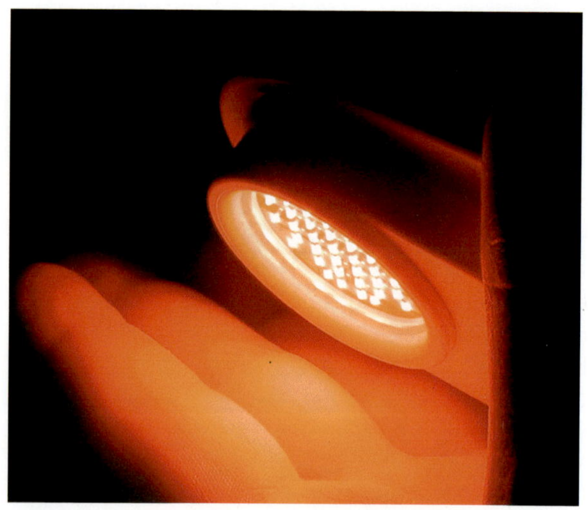

Red light-emitting diodes are growing plants in space and healing humans on Earth.

In addition to promoting cell growth, the red LEDs are capable of activating light-sensitive, tumor-treating drugs that, when injected intravenously, could completely destroy cancer cells while leaving surrounding tissue virtually untouched. The technique, approved by the U.S. Food and Drug Administration (FDA) for use in laboratory and human trials, is known as Photodynamic Therapy.

With the SBIR assistance from NASA, QDI set out to alter a surgical probe that could emit long waves of red light to stimulate a Benzoporphyrin-derivative drug called Photofrin, which delivers fewer post-operative side effects than comparable drugs. Ignatius additionally developed a friendly and successful working relationship with Dr. Harry Whelan, a professor of pediatric neurology and director of hyperbaric medicine at the Medical College of Wisconsin in Milwaukee. The two had met after Ignatius came across a newspaper article highlighting Whelan's ground-breaking brain cancer surgery technique, which uses drugs stimulated by laser lights to accelerate healing. Accordingly, QDI provided more than $1.25 million from its SBIR contracts to support Whelan's pioneering photobiomodulation research and bring him onboard to help improve the surgical probe.

Collectively, Ignatius, Whelan, and researchers from NASA successfully altered the probe for pediatric brain tumors and the prevention of oral mucositis (a common side effect of chemotherapy and radiation treatments) in pediatric bone marrow transplant patients at the Medical College of Wisconsin. In May 1998, a 20-year-old female became the first patient to undergo surgery with the modified probe. The young woman had endured six brain surgeries and chemotherapy and radiation treatments over a span of 10 years, but her aggressive cancer kept coming back. Having exhausted all of her conventional treatment options, she turned to the NASA-sponsored Photodynamic Therapy technology.

During the procedure, surgeons excised as much of the recurring brain tumor as they could, then injected the light-activated Photofrin into her bloodstream and inserted the LED probe into the remaining tumor tissue. The probe, which casts long wavelengths that generate less heat and penetrate deeper into tissue than the shorter wavelengths of traditional medical lasers, proved to be both safe and effective, as the tumor never returned and the patient recovered with no complications. A second operation that took place 3 months later on a male patient was also deemed successful by Whelan and his team of Medical College of Wisconsin surgeons.

FDA-approved clinical trials continued at several other facilities over the next 3 years, including the Roswell Park Cancer Institute in Buffalo, New York; Rush-Presbyterian-St. Luke's Medical Center in Chicago; and the Instituto de Oncologia Pediatrica in Sao Paulo, Brazil. QDI became recognized as a U.S. Space Foundation "Space Technology Hall of Fame" award recipient in 2000 and a Marshall Space Flight Center "Hallmark of Success" in 2004.

Product Outcome

The positive clinical trial results, as well as continued support from NASA and follow-on research grants from the Defense Advanced Research Projects Agency, helped QDI and the Medical College of Wisconsin to fully transition space technology into a new, non-invasive medical device. The WARP 10 (Warfighter Accelerated Recovery by Photobiomodulation) is a high-intensity, hand-held, portable LED unit intended for the temporary relief of minor muscle and joint pain, arthritis, stiffness, and muscle spasms. It also promotes relaxation of muscle tissue and increases local blood circulation. Unlike the surgical probe, the WARP 10 does not require intravenous medicine; instead, the unit can be placed directly on the skin where treatment is to occur.

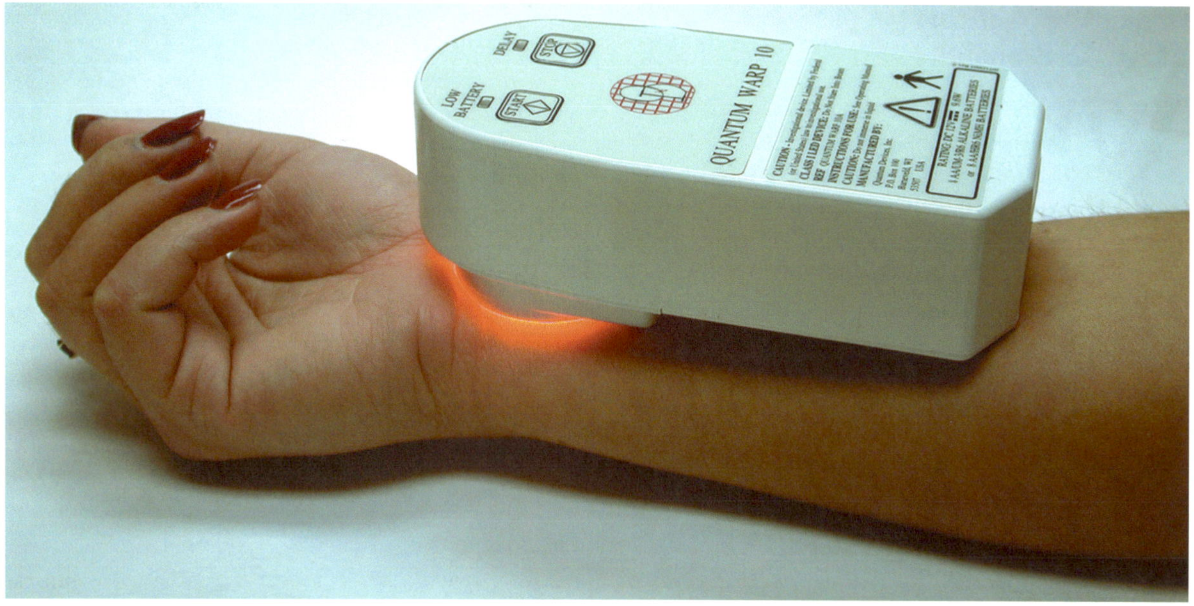

The WARP 10 was designed to assist armed forces personnel with immediate first aid and care for minor injuries and pain. A consumer version is also available.

The WARP 10 was designed to aid armed forces personnel on the front lines with immediate first aid care for minor injuries and pain, thereby improving endurance in combat. The "soldier self care" device produces 80 times more photon energy than a 250-Watt heat lamp, yet it remains cool to the touch. The power advantage reduces the time required for each therapeutic dose and provides for faster multi-dose exposures when needed, without the harmful effects of ultraviolet solar radiation. The U.S. Department of Defense and the U.S. Navy are currently issuing WARP 10 to crews on submarines and Special Forces operations.

QDI has introduced an FDA-approved consumer version sharing the same power and properties of the military model, as an alternative to the cost and complications associated with overuse of non-steroidal anti-inflammatory drugs (NSAIDs) for persistent pain relief. According to a Mayo Clinic study, adverse events associated with the use of NSAIDs are reported more frequently to the FDA than such events associated with any other group of drugs. Furthermore, conservative calculations for the United States estimate that approximately 107,000 patients are hospitalized each year for NSAID-related gastrointestinal complications and at least 16,500 NSAID-related deaths occur annually among arthritis patients alone, according to compiled research.

QDI is shedding the risks and costs linked with these anti-inflammatory drugs in favor of shedding light on safe and economical human healing. ❖

HEALS® and WARP 10® are registered trademarks of QDI, Inc.
Photodynamic™ Therapy is a trademark of QDI, Inc.
Photofrin® is a registered trademark of Axcan Pharma Inc.

Discovering New Drugs on the Cellular Level

Originating Technology/ NASA Contribution

With the Vision for Space Exploration calling for a sustained human presence in space, astronauts will need to grow plants, while in orbit, for nourishment that they will not receive from only consuming dehydrated foods. As a potential source of food for long-duration missions, space-grown plants could also give astronauts an important psychological boost, as fresh vegetables could serve as a welcomed change from monotonous meals consisting of reconstituted foods in plastic bags. Even more, these plants could likely aid in the recycling of air and wastewater on spacecraft.

With a helping hand from a company by the name of Biolog, Inc., NASA is studying the impacts of decreased gravity and spaceborne bacteria on the plants being grown for food in space.

With a helping hand from NASA, this very same company is creating powerful new cell- and bacteria-analysis tools for use in discovering and developing new drugs on Earth.

Partnership

From 1993 to 1997, Hayward, California-based Biolog received Phase I and II **Small Business Innovation Research (SBIR)** contracts from Kennedy Space Center to work with NASA in developing two technologies that are now in use by Biolog customers worldwide. The first technology, based on the company's pre-existing assay kits, was a process created to monitor populations of microbes. Dr. Jay Garland, manager of the Life Support Group at Kennedy, was interested in using this process as a way of monitoring the health of hydroponic crops that would be grown in space during future manned missions to Mars. The second technology was an instrumented system that would allow automated monitoring of Biolog's assay kits for NASA's purposes. Together, these innovations intended to provide NASA with better means of growing food in space and avoiding a catastrophic crop failure during long-term space travel.

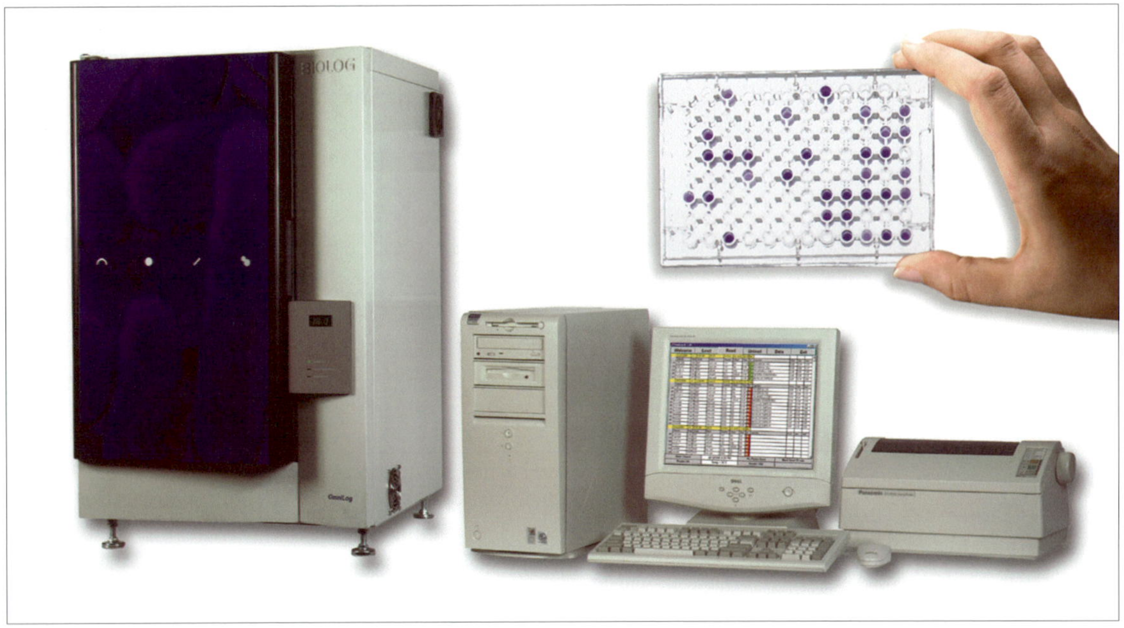

Biolog, Inc.'s product lines have been built upon patented technology that greatly simplifies testing of cells. Principal customers include pharmaceutical, biotech, and cosmetic laboratories, as well as laboratories testing for human, animal, and plant diseases.

Product Outcome

Biolog went on to extend the technologies it developed with SBIR assistance from NASA in order to bring two novel and important cell-testing technologies to market: the Phenotype MicroArray and the OmniLog System.

Biolog's Phenotype MicroArray technology is complementary to DNA microarrays and proteomic technologies, which allow scientists to detect changes in levels of genes or proteins that direct most cellular functions. By measuring the patterns of change in genes and proteins, scientists are attempting to correlate the findings with something important, such as a disease state. A major goal would be to understand the biochemical basis underlying a disease and gain insight into how to correct the problem. However, according to Biolog, there are typically hundreds to thousands of changes that are detected by these gene and protein analyses and it is often difficult to judge which ones are really significant to the cell.

"A big advantage we have with our Phenotype MicroArray cellular assays is that we are measuring change at the cellular level," claimed Dr. Barry Bochner, Biolog chairman and chief scientific officer. "Proteomics and gene expression don't necessarily give enough insight—often, you don't really know what it means to the cell. The data we get is at a higher level in terms of information content and can be simpler to interpret."

Because of this, the Phenotype MicroArray represents the third major technology that is needed in the genomic era of research and drug development, according to the

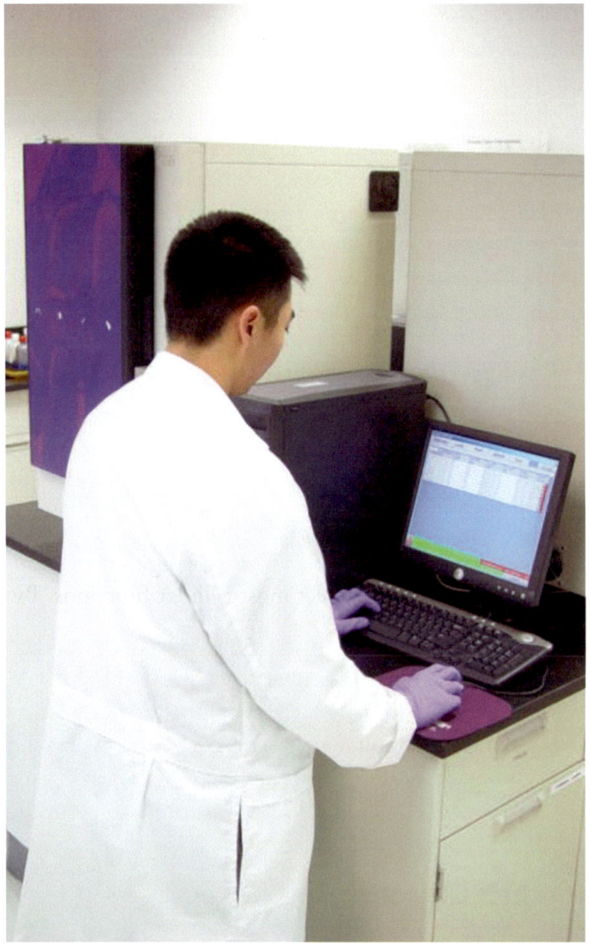

A Biolog, Inc., laboratory technician performs cellular tests using Phenotype MicroArray assays and the OmniLog System. The technology is capable of monitoring thousands of phenotypes simultaneously.

company. Just as DNA microarrays and proteomic science have made it possible to assay the level of thousands of genes or proteins all at once, the Phenotype MicroArray technology makes it possible to quantitatively measure thousands of cellular traits (phenotypes) simultaneously, in both microbial and mammalian cells. Such measurements can be critical in determining the effects of genetic changes and drug exposure on cells.

In the field of drug discovery, the Phenotype MicroArray allows researchers to obtain a comprehensive picture of a drug's effect on a specific cell. "People usually do this type of assay on cells in a rapid growth state, and they're only looking at the cell under one state," Bochner noted. "But a cell is always changing. With our technology, you can take a drug and get a very information-rich fingerprint of its effect on the cell under a wide range of physiological states."

Incubation and recording of phenotypic data gathered from the Phenotype MicroArray are performed by the OmniLog PM System, an integrated system of cellular assays, instrumentation, and bioinformatics software. Just as it automatically monitored assay kits for NASA, the OmniLog technology monitors thousands of phenotypes simultaneously. Several times each hour, it captures digital images of the cell assays being studied and stores quantitative color change values as computer files. These files can be displayed in the form of kinetic graphs and up to 480,000 data points can be generated in a 24-hour period.

Another version of the OmniLog product, called the OmniLog ID System, can be paired with Biolog's microbial-identification kits to easily and efficiently identify over 2,000 species of aerobic and anaerobic bacteria, yeasts, and fungi, virtually everything from A to Z (*Achromobacter cholinophagum* to *Zygosporium mycophilum*).

The principal customers for Biolog's Phenotype MicroArray and OmniLog products are laboratories requiring state-of-the-art capabilities in cell-based assay and identification products, especially pharmaceutical, biotech, cosmetics, and medical device companies; university and government research laboratories; laboratories testing for diseases of humans, animals, and plants; laboratories performing environmental monitoring; and companies or organizations involved in production or testing of food and drink.

Applications of the NASA-funded technology continue to expand and evolve. Biolog recently announced that the Phenotype MicroArray and OmniLog products have been installed at the Lawrence Livermore National Laboratory, in Livermore, California, where genomics researchers are using the technology to understand and characterize phenotypes of bacteria strains that are considered potential bioterrorism agents. Other important government laboratories such as the U.S. Food and Drug Administration and the U.S. Department of Agriculture are also employing the technologies to better understand foodborne pathogenic bacteria and the spread of epidemics.

Additionally, Biolog is sharing its innovations with the Nara Institute of Science and Technology, in Japan, in an effort to broaden what is known about the functions of genes in the important bacterium *E. coli*. According to Tim Mullane, Biolog president and chief executive officer, "*E. coli* is one of the most studied model cell lines in the world, and despite the early sequencing of this organism's genome, many of its 4,000 genes have no known function. Even where function is known, this information is often incomplete."

Next on the list for Biolog is to broaden the technologies for use with human cells. The company will soon be releasing Phenotype MicroArrays that can assess the energy metabolism pathways in cells from different organs and tissues. This should aid studies in diabetes, obesity, and cancer. ❖

Hydrogen Sensors Boost Hybrids; Today's Models Losing Gas?

Originating Technology/ NASA Contribution

Advanced chemical sensors are used in aeronautic and space applications to provide safety monitoring, emission monitoring, and fire detection. In order to fully do their jobs, these sensors must be able to operate in a range of environments. NASA has developed sensor technologies addressing these needs with the intent of improving safety, optimizing combustion efficiencies, and controlling emissions.

On the ground, the chemical sensors were developed by NASA engineers to detect potential hydrogen leaks during Space Shuttle launch operations. The Space Shuttle uses a combination of hydrogen and oxygen as fuel for its main engines. Liquid hydrogen is pumped to the external tank from a storage tank located several hundred feet away. Any hydrogen leak could potentially result in a hydrogen fire, which is invisible to the naked eye. It is important to detect the presence of a hydrogen fire in order to prevent a major accident.

In the air, the same hydrogen-leak dangers are present. Stress and temperature changes can cause tiny cracks or holes to form in the tubes that line the Space Shuttle's main engine nozzle. Such defects could allow the hydrogen that is pumped through the nozzle during firing to escape.

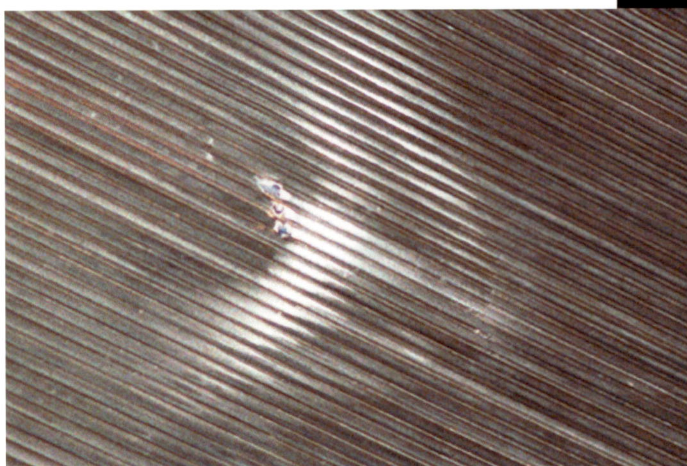

The image to the right shows what appear to be three small holes in the liquid hydrogen tubes inside the nozzle on Space Shuttle Main Engine No. 3 (right-most engine), following the landing of Space Shuttle Columbia (STS-93) on July 27, 1999. The smaller image above shows the holes in greater detail. During the mission, Columbia reached an orbit about 7 miles lower than anticipated, due to premature main engine cutoff that was traced to a hydrogen leak.

Responding to the challenges associated with pinpointing hydrogen leaks, NASA endeavored to improve propellant leak-detection capabilities during assembly, pre-launch operations, and flight. The objective was to reduce the operational cost of assembling and maintaining hydrogen delivery systems with automated detection systems. In particular, efforts have been focused on developing an automated hydrogen leak-detection system using multiple, networked hydrogen sensors that are operable in harsh conditions.

Partnership

In 1999, Glenn Research Center's Technology Commercialization Office awarded Makel Engineering, Inc., with a **Small Business Technology Transfer (STTR)** contract and additional funding to commercialize NASA's automated hydrogen-sensing technology for aerospace, industry, and consumer applications. Makel Engineering, based in Chico, California, worked closely with Glenn throughout the technology's commercial development. Recognizing that the NASA sensors could expedite the time-to-market for hydrogen-fueled transportation vehicles, the company went on to partner with a top U.S. automaker and supply its advanced hydrogen sensors for hydrogen-powered internal combustion engine (H2ICE) applications.

Product Outcome

The U.S. Government, auto manufacturers, and citizens have all encouraged the use of hydrogen as a transportation fuel. Transitioning to hydrogen would provide several advantages, such as reducing dependence on foreign oil and eliminating vehicle emissions. Before this future is realized, however, two essential principles must be addressed: the need to responsibly tackle overarching fuel-safety concerns and the need for fast, reliable hydrogen monitoring—integrating data collection, analysis, and communication.

Makel Engineering, Inc.'s advanced hydrogen sensors are built to operate in harsh conditions, for aerospace, industrial, and commercial applications.

Consequently, Makel Engineering is providing Ford Motor Company with hydrogen leak-sensing systems for its prototype H2ICE vehicles. The systems consist of four hydrogen sensors, a control unit, and associated cabling. The sensors are installed at various locations throughout the vehicle and continuously monitored by the control unit. In the event of a hydrogen gas leak, the sensors will alert the control unit to the presence of hydrogen, and the control unit can then initiate appropriate actions.

Ford regards the H2ICE as a near-term, low-cost transition or "bridging" strategy to stimulate the development and maturation of the hydrogen infrastructure and related hydrogen technologies, including on-board hydrogen fuel storage, hydrogen fuel dispensing, and hydrogen safety sensors. The engine has a laundry list of benefits that rival its gasoline-powered predecessor. It possesses all-weather capability with no cold-start issues, and requires zero warm-up. It has high efficiency (52-percent peak-indicated efficiency), as the vehicles it operates can easily achieve California's Super-Ultra-Low-Emission Vehicle (SULEV) standards and more than 99-percent reduced carbon dioxide vehicle emissions. Even more, its performance—while running comparable to gasoline—increases fuel economy by 25 percent, and up to 50 percent with an aggressive hybrid electric strategy.

The prototype version of the hydrogen-hybrid powertrain was unveiled to the public in January 2003, in the Model U concept vehicle at the North American International Auto Show (NAIAS) in Detroit. There, Ford touted the Model U as "a model for change for the next century" and "the Model T of the 21st century."

The drivable version of the supercharged, intercooled hydrogen powertrain was unveiled during the Ford Centennial celebration in June 2003. Scores of journalists from around the world were able to experience driving the H2ICE-equipped prototype vehicle firsthand during a media drive held in a Detroit-area park.

Ford is moving to put the hydrogen-powered technology to work in a V-10 shuttle bus engine, as well. The H2ICE E-450 chassis cab made its debut earlier this year at the 2005 NAIAS. The E-450 shuttle bus seats up to 12 passengers, including the driver, with room for luggage. The vehicle is equipped with a 5,000-psi hydrogen fuel tank and emits only water as exhaust. The automaker expects the shuttle bus to have a driving range of up to 150 miles, depending on conditions and vehicle load.

Makel Engineering notes that, as the use of hydrogen as a transportation fuel becomes more prevalent, numerous technological solutions for hydrogen generation, storage, and utilization will be created—all having stringent safety requirements. Furthermore, it asserts that, as hydrogen becomes a more practical and established fuel source, the availability of safe hydrogen refueling sources will be fundamental to public acceptance. Effective hydrogen sensors that respond accurately and quickly to hydrogen gas leaks will be a prerequisite in the development of hydrogen-powered vehicles and related infrastructure.

Makel Engineering's development and delivery of a vehicle-safety sensor system demonstrates acceptable levels of performance, reliability, and cost, and overcomes a major commercialization barrier for transportation applications. ❖

Makel Engineering, Inc., worked with Ford Motor Company's research and development team on a comprehensive hydrogen-monitoring system for the Model U prototype. The system provides continuous leak monitoring throughout the vehicle and has been demonstrated at car shows and other advanced automotive technology events.

3-D Highway in the Sky

Originating Technology/ NASA Contribution

If it were 50 years ago, NASA's contribution to rock and roll could have been more than just the all-astronaut rock band, Max Q, composed of six NASA astronauts, all of whom have flown aboard the Space Shuttle. If it were 50 years ago, a new NASA spinoff technology, Synthetic Vision, would likely have been able to prevent the fateful, small plane crash that killed rock and roll legends Buddy Holly, Ritchie Valens, and The Big Bopper on that stormy night in 1959. Synthetic Vision is a new cockpit display system that helps pilots fly through bad weather, and it has incredible life-saving potential.

In 1997, the White House Commission on Aviation Safety and Security created NASA's Aviation Safety and Security Program (AvSSP) with the aim of sounding the depths of NASA's cutting-edge aviation advances and history of successes. The AvSSP decided to use NASA technology to cut the rate of fatal aviation accidents that occur because of lowered visibility and spatial disorientation, common problems that arise in poor weather conditions, in the dark, or with inexperienced pilots.

Partnership

Aeronautics researchers at NASA's Langley Research Center teamed with Chelton Flight Systems and the Federal Aviation Adminstration (FAA) Alaskan Region Capstone Program, a technology-focused safety program which seeks safety and efficiency gains in aviation by accelerating implementation of modern technology.

Because of the harsh Alaskan terrain, planes in this region are used for most common errands, whether it is a trip to the dentist or a grocery and supply run. Aviators in these unforgiving climates often fly low, around 200 feet above the ground, to avoid poor visibility and icing conditions in winter, and rely on lighted buoys anchored in lakes to guide their transit. It is the ideal area in which to test weatherproof guidance symbology. As part of the Synthetic Vision research and development testing conducted at Langley, a display concept replicating the Chelton display system was included in the test matrix and evaluated in simulation and flight-test experiments.

Synthetic Vision flight displays present real-time depictions of the terrain and complement the view the pilot has through the cockpit windshield. Here it has been installed in a Beechcraft King Air 300.

The result of the research is a 3-D display for pilots, which provides clear vision, regardless of the time of day or weather conditions. It also replaces the buoys with a series of onscreen markers that draw a virtual highway in the sky, on which the pilots can "drive."

The simulated tests conducted by NASA, the FAA, and Chelton involved dozens of pilots in a variety of scenarios, with one specifically designed to cause an accident. The testing confirmed that with the Synthetic Vision system in place, fewer accidents occurred. In fact, Synthetic Vision lowered the chance of hazardous events 85 percent over traditional instrumentation. Because of these stellar results and continuing positive feedback from test pilots, Chelton's Synthetic Vision system earned its name and its place as the first commercially available system of its kind ever offered. It is already being marketed and distributed by Chelton for use on small planes, light jets, and helicopters. The system is flying in hundreds of small planes all over the United States and abroad.

As a result of this successful and beneficial partnership, Langley, the FAA, and Chelton were presented with NASA's "Turning Goals into Reality" award for aviation safety.

Product Outcome

Synthetic Vision allows pilots to see their surroundings as if the sky were clear, regardless of the actual environmental conditions. It presents a real-time, forward-looking depiction of the terrain in 3-D, directly on the primary flight display. The pilot has a virtual view of his surroundings, in addition to the view afforded by the cockpit windshield. The system creates an artificial, computer-generated view of the surroundings from a series of databases and advanced sensory input, so it gives pilots topographical flight plans, as well as real-time feedback, about the area directly outside of the aircraft.

It is, at its core, a database-driven system, using onboard terrain, obstacle, and airport information databases and employs a highly precise navigational system to position the aircraft within those databases' parameters. The system also contains a number of database integrity-monitoring technologies to help ensure that the presentation given to the flight crew is indeed a correct one. The pilot can program a flight plan into the Synthetic Vision system and the onboard computer will know which databases to access, and then provide the pilot with course markers that highlight the route the plane should take. These markers can even extend into the landing zone and create a trail that the pilot can follow straight through to landing.

In addition to the databases, Synthetic Vision may employ a series of advanced sensors that are sensitive enough to identify objects within close proximity of the aircraft. It extends the basic capabilities typically found on flight systems and, rather than just showing large topographical features like mountains, ridges, and valleys, it extends to man-made structures like buildings, towers, and other obstructions, such as vehicles on a runway.

The system is advanced to the degree that it can point out variations previously overlooked by traditional global positioning systems (GPS). For example, if a mountain ridge were to rise above the horizon, the Synthetic Vision system could alert the pilot; whereas, with typical GPS, the pilot would merely know that a mountain was near and the average altitude of the mountain. Synthetic Vision alerts the pilot with an audio signal to change the course of the airplane in order to avoid collision.

Not only does the Synthetic Vision system provide more information to the pilot, it is also easier to read. It flows naturally, as opposed to being broken into choppy video on the heavily-pixilated screens typical of most in-flight displays. The full-color, high-resolution screens mount in place of the cockpit flight displays and provide smooth, streaming video representations of what is happening in front of the plane.

Although successfully used in the Alaskan Capstone project and throughout the "lower-48," it will still take time before the Synthetic Vision system is made available to commercial airlines. The Chelton EFIS Primary Flight Display has achieved Level A, the highest level of software standards established by the Radio Technical Commission for Aeronautics for the FAA. This standard is essential for all flight-critical avionics. The FAA is strict about approving new technologies for commercial airline use, and it is expected that this acceptance, though likely in the future, will occur after additional years of testing the technology. In addition, though Synthetic Vision has demonstrated its safety benefits, the airlines require an efficiency benefit before expending the capital to bring the technology aboard the aircraft. Subsequent Langley research is developing and evaluating these potential benefits. ❖

EFIS Primary Flight Display™ is a trademark of Chelton Flight Systems.

This top-down indicator of a flight approach to the Juneau International Airport in Alaska shows wind information, air temperature, true airspeed, and groundspeed, as well as the plane's relationship to any towers, antennas, or obstructions, and additional information pilots use to fly safely in this arctic climate.

Popping a Hole in High-Speed Pursuits

Originating Technology/ NASA Contribution

NASA's Plum Brook Station, a 6,400-acre, remote test installation site for Glenn Research Center, houses unique, world-class test facilities, including the world's largest space environment simulation chamber and the world's only laboratory capable of full-scale rocket engine firings and launch vehicle system level tests at high-altitude conditions. Plum Brook Station performs complex and innovative ground tests for the U.S. Government (civilian and military), the international aerospace community, as well as the private sector.

Recently, Plum Brook Station's test facilities and NASA's engineering experience were combined to improve a family of tire deflating devices (TDDs) that helps law enforcement agents safely, simply, and successfully stop fleeing vehicles in high-speed pursuits.

Partnership

Phoenix International, Ltd., of Brookfield, Wisconsin, has been manufacturing and marketing the MagnumSpike! spike-lined TDDs to law enforcement since 1986. Up until the company's involvement with Plum Brook Station, spikes were inadvertently being knocked onto the roadways whenever tires from fleeing cars and trucks were deflated by the device. This was caused by the enormous impact created when vehicles are brought to a halt. Cleaning up these spikes from out of the roadway was problematic for officers.

Phoenix International needed a mechanism to keep the spikes in place until they are ready for release. The small, woman-owned company learned about NASA's Garrett Morgan Commercialization Initiative, a program operated by Glenn that allows minority companies to grow and strengthen their businesses by leveraging NASA technology, expertise, and programs. Realizing the valuable assistance it could provide in improving the MagnumSpike!, the Garrett Morgan Commercialization Initiative paired Phoenix International with Glenn engineer Chip Redding.

Redding evaluated the MagnumSpike! in order to understand and control the variables preventing repeatable, reliable performance. His analysis was then transferred to a Cleveland-based engineering firm that conducted additional tests for precise measurements of the retention system used to lock in the spikes. Based on the analyses, both Redding and the engineering firm provided Phoenix International's engineers with recommendations on alterations to the mechanism that can provide a precise, friction-free release of the spikes.

Additionally, the MagnumSpike! was tested at Plum Brook Station, where researchers concluded that its retention clip could not reliably hold all of the device's spikes in place. As a direct result of this assistance, Phoenix International decided to develop a pin shear method to incorporate into its line of MagnumSpike! TDDs. Glenn and the Garrett Morgan Commercialization Initiative granted Phoenix International a financial award, through a Simplified Technology Transfer Agreement, to assist with completion of the design, testing, and manufacturing of new, friction-free retain-release clips that were generated from the new pin shear method.

The financial award was also used for product marketing and promotion of the device, in addition to further

The MagnumSpike! tire deflation device is designed to safely, swiftly, and consistently end high-speed pursuits within a predictable short distance. It gives law enforcement officers control over extremely dangerous situations, preventing property damage and safeguarding not only their lives, but the lives of innocent bystanders and the occupants of the pursued vehicles.

One of hundreds of trucks stopped by MagnumSpike!

development and initial manufacturing of a new vehicle stopping product called the StaticStop, which consists of a disposable spiked wheel chuck that can be used to detain vehicles in routine traffic stops.

"R&D is the lifeblood of what we do," noted Toranj Marphetia, Phoenix International's president and chief executive officer. "The Garrett Morgan Commercialization Initiative provided us with much-needed expertise to evaluate and measure our new [MagnumSpike!] mechanism," she added.

Product Outcome

The new, improved MagnumSpike! has been field tested and "pursuit proven" as the ultimate weapon against high-speed chases. It swiftly and safely stops everything from 18-wheelers to compact vehicles, even vehicles equipped with new self-sealing and run-flat tires. (According to tests conducted by the National Institute of Justice—the research, development, and evaluation agency of the U.S. Department of Justice, the MagnumSpike! is the only TDD on the market to consistently stop vehicles outfitted with these new types of tires.) Decreasing chase time for police forces increases officer safety, saves lives, and prevents property damage.

"We have used the MagnumSpike! at least 30 times with complete success in ending the pursuit without injuries or cars getting damaged," said Gary Miller, sheriff of Wright County, Minnesota. "From our testing and experience, I can say the MagnumSpike! is the best. It does the job every time, regardless of vehicle size," he added.

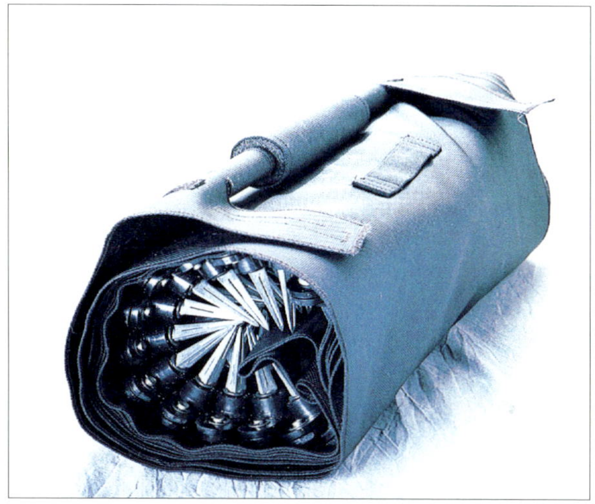

The "wrap and roll" MagnumSpike! model, packed in a soft ballistic nylon case, is 12 feet in length, weighs 9 pounds, and contains 160 spikes.

Available several sizes and models, the MagnumSpike! can have up to 250, 2.5-inch, precision-cast, rust-proof spikes, able to withstand multiple high-speed impacts. Each spike contains three sharp corners. Additionally, a patented choke-proof design prevents them from getting stuck in the deflated tires.

The technology, which also has application in U.S. Department of Defense and U.S. Department of Homeland Security operations, is currently flaunting an impressive 100-percent safety record and 100-percent take-down record in deflating all tires that have come across its path. In one unique instance, the MagnumSpike! made an impact—without making impact. When a fleeing suspect saw the strip of spikes, he skidded his vehicle to a stop, got out, and surrendered.

Phoenix International anticipates that the improved MagnumSpike! will widen its market five-fold, leading to a major increase in sales—or, a major "spike" in business. ❖

MagnumSpike!™ is a trademark of Phoenix International, Ltd.

Monitoring Wake Vortices for More Efficient Airports

Originating Technology/ NASA Contribution

Wake vortices are generated by all aircraft during flight. The larger the aircraft, the stronger the wake, so the Federal Aviation Administration (FAA) separates aircraft to ensure wake turbulence has no effect on approaching aircraft. Currently, though, the time between planes is often larger than it needs to be for the wake to dissipate. This unnecessary gap translates into arrival and departure delays, but since the wakes are invisible, the delays are nearly inevitable.

If, however, the separation between aircraft can be reduced safely, then airport capacity can be increased without the high cost of additional runways. Scientists are currently studying these patterns to identify and introduce new procedures and technologies that safely increase airport capacity. NASA, always on the cutting edge of aerospace research, has been contributing knowledge and testing to these endeavors.

WindTracer detects wind hazards such as windshear, microbursts, gust fronts, turbulence, crosswinds, and wake vortices that can compromise the safety of an aircraft.

Partnership

NASA's Langley Research Center, working with the FAA on a joint program known as the Wake Turbulence Research Program, conducted research on the wake vortices at the Denver International Airport.

Langley scientists collected acoustic signature data from the wakes of landing aircraft and then characterized these signatures for a variety of aircraft types in various conditions. Two large, precision microphone arrays were placed on the ground beneath the glide slope for a runway, one operated by NASA and the other by the German Aerospace Center, or Deutschen Zentrum für Luft- und Raumfahrt. Because there are numerous unknowns about the acoustic signatures of wakes, a truth sensor was needed to tell NASA the location of the wake, allowing the measurements to be properly characterized.

The research teams used WindTracer as a ground truth sensor for these wake vortex acoustic tests. WindTracer is manufactured by Coherent Technologies, Inc. (CTI), the recipient of a Langley **Small Business Innovation Research (SBIR)** grant to develop this integral piece of equipment. CTI is a privately held company based in Louisville, Colorado, and is the only company in the world that offers an infrared Doppler lidar as a commercial product.

Product Outcome

WindTracer uses an infrared, eye-safe laser, with precision pointing and scanning capabilities, to bounce off the natural particulates floating in air. The light reflected back to the system measures the wind and tracks the strength and location of the aircraft vortices. The technology has been developed over the past decade by CTI and is applied to a variety of airport and airliner wind hazards, as well as measurement applications.

WindTracer detects wind hazards such as windshear, microbursts, gust fronts, turbulence, crosswinds, and wake vortices that can compromise the safety of an aircraft. It detects these hazards and transmits real-time data to air traffic control display monitors, providing immediate audio and visual alerts to the staff when dangerous conditions arise. This information can be quickly relayed to the pilots of arriving or departing aircraft who can then adjust their patterns to avoid the hazards.

A WindTracer unit is currently set up at the Hong Kong International Airport, which is located near Lantau Island, a large mountainous island that often experiences windshear and turbulence. Other wind hazards arise due to frontal passages and sea breezes. These wind conditions are potentially hazardous to landing and departing aircraft.

The device is operated by the Hong Kong Observatory (HKO), one of the world's leading meteorological organizations. It forecasts weather and issues warnings on weather-related hazards at the airport and within a designated airspace over the northern part of the South China Sea. To enhance the safety of aircraft landing at and taking off from the airport, the HKO issues alerts of low-level windshear and turbulence. A Terminal Doppler Weather Radar network of over 20 anemometers, 2 wind profilers, and WindTracer are used to assist in the detection and warning of windshear and turbulence.

It is also being employed successfully by the St. Louis International Airport, where it is used to observe wake vortices produced by aircraft landing on specific runways. It then provides data, which is used to validate the safety of new capacity-enhancing procedures being developed by the FAA.

WindTracer profiles winds and detects windshear, turbulence, and aircraft wake vortices at both airports and onboard commercial airliners.

Since its installation in 2003, the system has been running unattended, with system operational modes being automatically changed via an operator-defined schedule, and remote access to the system via an Ethernet connection enables mode and schedule changes to be affected without the need for onsite staff.

The WindTracer was also used for the U.S. Department of Defense's (DOD) atmospheric dispersion survey conducted in and around the Pentagon. The survey sought to improve knowledge about the weather conditions and movement of simulated airborne contaminants. Knowledge gained about the airflow around the Pentagon, and the associated transport of gases and their infiltration into the building, will lead the development for improved systems to protect other DOD facilities.

The product has been proving itself useful around the world, and it has the potential to revolutionize the entire airline industry. ❖

WindTracer uses an infrared, eye-safe laser, with precision pointing and scanning capabilities, to bounce off the air particulates. It poses no risks to pilots.

WindTracer® is a registered trademark of Coherent Technologies, Inc.

From Rockets to Racecars

Originating Technology/ NASA Contribution

NASA's Langley Research Center scientists developed a family of catalysts for low-temperature oxidation of carbon monoxide and other gases. The catalysts provide oxidation of both carbon monoxide and formaldehyde at room temperature without requiring any energy input, just a suitable flow of gas through them.

Originally designed as part of an atmospheric satellite project, where the catalysts were intended to recycle and recapture carbon dioxide to enhance the operational life of carbon dioxide lasers, the entire system was made to be rugged, long-lived, and fail-safe.

The low-temperature oxidation catalysts can be produced and coated onto various catalyst supports, including porous ceramic monoliths and beads, which means that they can be integrated into existing designs, made to fit in limited space, and fashioned into a variety of geometrically different products.

Although the satellite project was never launched, the resulting catalysts are doing great things here on Earth, with current applications in the high-speed motor sports arena as air purifiers, so professional racecar drivers do not get carbon monoxide poisoning. Future benefits may extend even further.

Partnership

The aerospace industry has always had a direct link to the automotive industry, and NASA has, over the years, provided a great deal of its technology to its ground-based cousin, whether it was in the form of grooved pavement to reduce the risk of hydroplaning vehicles, or helping to design crash test dummies with embedded sensors. NASA has also contributed to automobile industry technological advances such as a software program that measures tire safety and fuel cell research that is revolutionizing the next generation of hybrid vehicles.

NASA has close ties with the high-performance competitive driving circuit. The Agency's latest contribution to the sport comes in the form of an air catalyst—part of a filter to keep drivers safe from the carbon monoxide lingering over the track.

NASA's operations also have something in common with the rigors of high-performance competitive driving. Both require reliability under extreme conditions with little or no maintenance. Add to that the fact that NASA currently holds several Guinness Book of World Records records for speed, and the connection becomes even clearer. The high-speed motor sports industry has benefited from NASA in several ways. Under a Space Act Agreement between Boeing North America, Penske Racing, and BSR Products, Space Shuttle Thermal Protection System materials are now used to insulate racecars. In the early days of the Space Shuttle Program, NASA scientists and the 3M company worked to improve high-temperature tiles and textiles to withstand the intense heat and pressures of reentry. These tiles, now manufactured by the 3M Company, are used on the floorboards of National Association for Stock Car Auto Racing (NASCAR) vehicles to block engine heat. These two innovations keep drivers safe when they are zooming by at speeds in excess of 200 miles per hour.

Additionally, the large 30- by 60-foot wind tunnel at Langley, currently leased to Old Dominion University, is being used to test airflow over racecars. Future plans are in the works to perform even more racecar testing at Langley.

The newest addition to NASA's contributions to high-speed motor sports is the new family of air catalysts. Penske Racing, of Mooresville, North Carolina, has designed an air filter using the NASA catalysts for drivers on the high-speed racing scene. STC Catalysts, Inc. (SCi), a subsidiary of Science and Technology Corporation, in

Hampton, Virginia, manufactures the catalysts and has been supplying them to Kustom Komponents, of Temple, Pennsylvania, for use in these filters.

Product Outcome

The filtration unit being sold to racecar drivers is about the size and shape of a Thermos bottle. It is part of a compound filter system containing absorbents and other materials to effectively remove noxious gasses from the driver's air stream. In order to be implemented into the design of the high-performance cars, it has to be small and compact so as not to add unwanted weight. Its capabilities derive from a unique surface chemistry and airflow over and through a formulation of platinum and tin oxide incorporated within a honeycombed form. As fumes enter the passenger-side air duct, they flow through an activated carbon filter, the carbon monoxide scrubbing catalyst, and finally through a 99.997-percent high efficiency particulate air (HEPA) filter. This unit filters out harmful gases before they affect the driver's health. The air is then cooled before being delivered directly to a port in the driver's helmet.

Cleaner air virtually eliminates carbon monoxide poisoning and the resulting flu-like symptoms, like headaches, fatigue, and dizziness that have traditionally lingered for days after races. The symptoms, compared by drivers to the worst hangover in the world, come from prolonged exposure to carbon monoxide as the drivers circle the track in a pack. Carbon monoxide poisoning occurs when so much of it is inhaled that it starts to replace the oxygen that is carried in the blood. Carbon monoxide binds to red blood cells about 250 times more strongly than oxygen. As the oxygen is replaced by the carbon monoxide, the body's organs and tissues cannot work properly.

The racecars are so close to each other and run so many laps, that the drivers literally never have a break from the carbon monoxide. Additionally, racecars do not have catalytic converters to reduce carbon monoxide emissions like commuter cars do, and they produce more emissions per hour due to the high speeds they travel. Exposure to the elevated level of fumes causes drivers to feel woozy, nauseous, and fatigued. It has even led to early retirement for some.

Rick Mast, a veteran racecar driver who boasted 364 starts over the course of his 15-year career, retired in 2003, citing chronic carbon monoxide poisoning as the cause. He complained of dizziness, headaches, and nausea that eventually ended his career. Other drivers spoke out about health problems resulting from carbon monoxide inhalation, and NASCAR responded by initiating a series of tests and looking for a filtration system that would keep drivers safe, something it called its "fresh air" study.

The clean air pumped through this filter virtually eliminates carbon monoxide poisoning and the resulting flu-like symptoms, like headaches, fatigue, and dizziness that have traditionally lingered for days after races.

Penske Engineering, under the direction of its president, Don Miller, drove the effort to design, develop, and demonstrate the filter. Penske worked with researchers at one of Langley's wind tunnels to control and isolate cooled air for filtration. Traditional racecars filter air before it is cooled, and then use most of the cooled air for heat rejection. Penske, however, through its work at the wind tunnel, managed to isolate an air stream, cool it, and then, using the precision air flow, design the filter.

Dubbed the "INCAR System" and manufactured and marketed by Kustom Komponents, approximately 190 kits have been produced and are in use by major racing teams, including Penske Racing, Roush Racing, Hendrix Motorsports, and Robert Yates Racing, among others. Estimates show that the reduction in harmful gasses is as much as 70 percent, or more, depending upon track and humidity conditions. This virtually eliminates the sickness drivers often refer to as getting "gassed." It is now common for drivers in high-speed motor sports to have the Penske filtration system installed in their cars.

Since the debut of its partnership with the racing industry, SCi's business has boomed. In fact, its needs have outgrown its building size. A move to more spacious quarters will be completed by 2006, and the company intends to upgrade its laboratory, add staff, and triple its catalyst production capacity.

Penske, on the other hand, is not profiting much from this venture economically; but rather, it invested the research and capital to protect drivers from carbon monoxide poisoning, and that, so far, has been a very satisfactory result. ❖

All-Terrain Intelligent Robot Braves Battlefront to Save Lives

Originating Technology/ NASA Contribution

As NASA's lead center for creating robotic spacecraft and rovers, the Jet Propulsion Laboratory (JPL) builds smart machines that can perform very complicated tasks—far, far away from the homeland.

JPL's robotic proficiency is making an impact millions of miles away on Mars, where two rovers are presently unlocking the secrets of the Red Planet's rugged terrain, and thousands of miles away in the embattled regions of Iraq and Afghanistan, where robots sown from the seeds of JPL machines have been deployed to be the "eyes and ears" of humans on the front line. This commercial offspring, known as the PackBot Tactical Mobile Robot, is manufactured by iRobot, Inc., of Burlington, Massachusetts.

Partnership

A number of iRobot employees have drawn from their NASA experiences to help develop PackBot, including the company's two co-founders. Colin Angle, the chief executive officer, designed behavior-controlled rovers for NASA that led to the Sojourner rover's exploration of Mars in 1997, and Helen Greiner, the president, worked at JPL as a student, building gripper systems for space satellites.

Additionally, Tim Ohm, one of iRobot's senior mechanical engineers, honed his technical expertise with NASA. While at JPL, Ohm helped build a NASA Martian rover whose structural features would ultimately be incorporated into PackBot. Nicknamed "Rocky-7," this Martian rover served as a terrestrial test bed for the twin Mars exploration rovers, Spirit and Opportunity, currently on Mars. Ohm worked on the lightweight, high-torque actuators used to control Rocky-7 and helped develop its strong, lightweight frame structure and its riveted sheet-metal chassis; the commercial PackBot robot has adopted these lightweight features and the same chassis concept.

In a separate effort, JPL's research facilities produced payloads for what was considered the official prototype to PackBot, "Urbie," while its Machine Vision group led the design and implementation processes, all through sponsorship from the Defense Advanced Research Projects Agency's (DARPA) Tactile Mobile Robot program. Urbie, short for urban robot, was a joint effort of DARPA, JPL, iRobot, the Robotics Institute of Carnegie Mellon University, and the University of Southern California Robotics Research Laboratory. The robot's initial purpose was mobile military reconnaissance in city terrain, but its rugged and lightweight features also made it useful for police, emergency, and rescue personnel.

With its rugged design, Urbie could be thrown into a building through a window, climb stairs, and fall from a balcony. These actions allow it to investigate urban environments contaminated with radiation, biological warfare, or chemical spills, or search earthquake-stricken buildings and other disaster zones.

Satisfied with Urbie's developmental progress, the DARPA-sponsored research consortium was ready to move forward with the field-ready successor: PackBot. iRobot was made the lead systems integrator and delivered the new PackBot tactical robots to the U.S. Army, the U.S. Air Force, and the U.S. Navy.

The resulting PackBot technology delivered benefits back to NASA, as well, considering PackBot's wheel design inspired the wheel design used for Spirit and Opportunity. In fact, the same machining shop that made the first PackBot wheels also made the Mars exploration rovers' wheels.

Product Outcome

PackBot's first military deployment was to Afghanistan in July 2002, where it was used by U.S. troops to help clear caves and bunkers, search buildings, and cross live anti-personnel mine fields. Prior to deployment, the Army Vice Chief of Staff, General Jack Keane (now retired), saw pictures of soldiers clearing caves with grappling hooks. Keane knew that the military had invested heavily in robotic equipment and did not see why the soldiers were still using this World War II-era technique. As a result, a "team" of PackBot robots was delivered to work with U.S. troops on a "send the robot in first" basis. The soldiers, who had never previously trained with the robots, were quickly won over.

An autonomous stairclimbing behavior is used to take NASA's "Urbie" robot up multiple flights of stairs without any user control. This is accomplished with a combination of onboard sensors and vision algorithms to sense where the stairs are and which direction the robot should go to drive up the center of the stairs.

Already proven in Afghanistan and Iraq, the PackBot Scout is the ideal "point man" to search dangerous or inaccessible areas, providing soldiers with a safe first look so they know what to expect and how to respond.

"PackBot is as tough, if not tougher, than any piece of military equipment I have ever used," according to one U.S. colonel stationed in Afghanistan.

The configuration sent to Afghanistan was the Scout, the simplest, lightest, and the most rugged and survivable PackBot model. The Scout weighs just under 40 pounds (18 kilograms) and is less than 8 inches (20 centimeters) high. It can be loaded into the Army-issued modular lightweight load-carrying equipment (MOLLE) backpack and deployed in minutes. This proved valuable for the Afghanistan missions, as troops frequently had to backpack the robots up the sides of mountains to reach cave entrances. Once deployed, the Scouts can quickly traverse narrow, difficult, hard-to-access terrain, such as caves, tunnels, and bunkers, and cover open ground at speeds up to 8.7 miles (14 kilometers) per hour. This model also possesses a fixed "reconnaissance" head that is equipped with multiple cameras; optional thermal imaging and infrared cameras can be used for night operations.

A second model, called the PackBot EOD, is being deployed in Iraq, in the ongoing fight against improvised explosive devices (IEDs). This robot replaces the need for a dangerous manual approach. Several systems have been damaged or completely destroyed in seeking out IEDs, but have been credited with saving lives in doing so.

Weighing less than 53 pounds (24 kilograms), the PackBot EOD can be carried by a single soldier and deployed in less than 2 minutes. It is equipped with a maneuverable arm that can reach as far as 6.6 feet (2 meters) in any direction to safely reach and disrupt difficult-to-access IEDs and landmines. The arm also allows PackBot to deploy tools, emplace counter charges, and recover forensic evidence. A stabilized telephoto zoom camera at the end of the arm provides assessment and surveillance at standoff distances.

A third model, the PackBot Explorer, boasts a continuous rotating pan-and-tilt head that can rise from the chassis and allow operators to peer over obstacles and gain greater perspective. The Explorer payload comes complete with multiple cameras, a laser pointer, and surveillance and Global Positioning System sensors.

For all PackBot models, operator ease-of-use is provided through intuitive onscreen displays and menus. In addition, the robots' intelligent behavior eliminates the drudgery of common tasks and protects against accidental misuse, according to iRobot.

PackBot's patented, self-righting mobility platform is outfitted with dual QuickFlip track articulations, so that the robot can climb grades up to 60 percent and survive submersion in water up to 6.6 feet (2 meters) deep. These "flippers" are capable of continuous, 360-degree rotation that propels the robot up stairs, over curbs, and through daunting obstacles such as rocks, rubble, and debris. If PackBot flips over during operations, the robot uses its flippers to perform a self-righting maneuver in seconds.

Already proven in two recent war scenarios, PackBot is the ideal "point man" to search dangerous or inaccessible areas, providing soldiers with a safe first look so they know what to expect and how to respond. ❖

PackBot® is a registered trademark of iRobot, Inc.
QuickFlip™ is a trademark of iRobot, Inc.

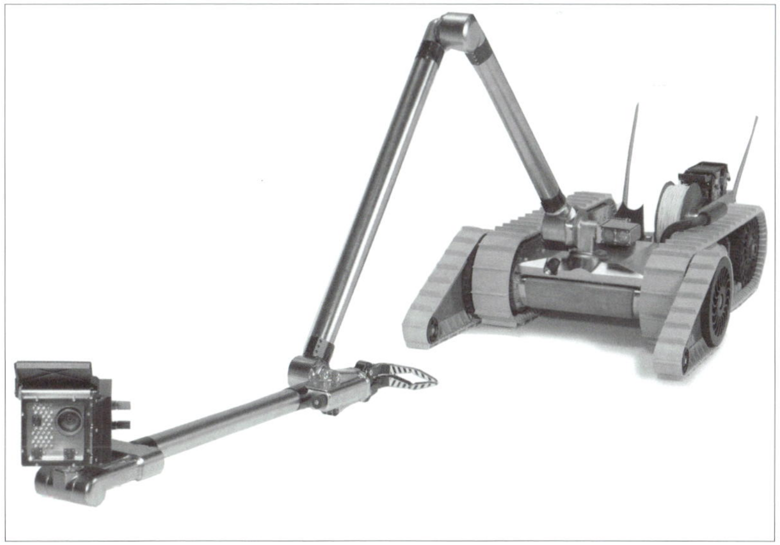

PackBot EOD is a rugged, lightweight robot designed to conduct explosive ordnance disposal, HAZMAT, search-and-surveillance, hostage rescue, and other vital law enforcement tasks for bomb squads, SWAT teams, military units, and other authorities seeking to meet the security challenges of the 21st century.

Keeping the Air Clean and Safe—An Anthrax Smoke Detector

Originating Technology/ NASA Contribution

Scientists at work in the Planetary Protection division at NASA's Jet Propulsion Laboratory (JPL) sterilize everything before blasting it to the Red Planet. They take great pains to ensure that all spacecraft are void of bacterial life, especially the microscopic bacteria that can live hundreds of years in their spore states. No one is quite sure what Earthly germs would do on Mars, but scientists agree that it is safest to keep the Martian terrain as undisturbed as possible. Errant Earth germs would also render useless the instruments placed on exploration rovers to look for signs of life, as the life that they registered would be life that came with them from Earth.

A team at JPL, headed by Dr. Adrian Ponce, developed a bacterial spore-detection system that uses a simple and robust chemical reaction that visually alerts Planetary Protection crews. It is a simple air filter that traps micron-sized bacterial spores and then submits them to the chemical reaction. When the solution is then viewed under an ultraviolet light, the mixture will glow green if it is contaminated by bacteria. Scientists can then return to the scrubbing and cleaning stages of the sterilization process to remove these harmful bacteria.

The detection system is the space-bound equivalent of having your hands checked for cleanliness before being allowed to the table; and although intended to keep terrestrial germs from space, this technology has awesome applications here on Mother Earth. The bacterial spore-detection unit can recognize anthrax and other harmful, spore-forming bacteria and alert people of the impending danger.

As evidenced in the anthrax mailings of fall 2001 in the United States, the first sign of anthrax exposure was when people experienced flu-like symptoms, which unfortunately, can take as much as a week to develop after contamination. Anthrax cost 5 people their lives and infected 19 others; and the threat of bioterrorism became a routine concern, with new threats popping up nearly everyday. The attacks threatened the safety that so many Americans took for granted, as the very air that people breathed became suspect. Any building with a circulation system, where large groups congregate, was now a potential target.

Ponce recognized a need for the application of his device and the timely terrestrial uses for this technology. "What we needed," he said, "was an automated air monitor that could warn us of an anthrax attack, much like a smoke detector warns us of a fire. This is exactly what we developed at JPL."

Universal Detection Technology's bacterial spore detection system, the BSM-2000, is simple and robust, prerequisites for continuous monitoring of the air.

The BSM-2000 is portable. The average unit is just about the size of a computer terminal and weighs less than 50 pounds.

Partnership

Realizing the enormous potential that the so-called anthrax "smoke" detector (ASD) held, NASA determined that the JPL venture could benefit from a commercial partner to speed this life-saving tool to market, where it could best help people.

The connection happened at a meeting on bioterrorism, hosted by the U.S. National Institutes of Health, where Greg Bearman, a scientist at JPL who worked with Ponce on the bacteria-testing devices and is active in JPL's Technology Transfer Program, happened to be seated next to Jacques Tizabi, president and CEO of Universal Detection Technology (UDT). In the air-sampling market for over 30 years, UDT found previous success with its monitors for ozone, nitrogen oxides, sulfur dioxide, and other pollutants. After a brief conversation, it was clear that the ASD was a perfect fit for the plans UDT had to improve its line of detection devices.

In August 2002, just months after the anthrax attacks shook the country, UDT partnered with JPL through NASA's Innovative Partnerships Program; and through the California Institute of Technology, it gained exclusive rights to option the ASD for worldwide use. This agreement also enabled UDT to leverage the expertise of JPL researchers, as they developed their bacterial spore-detection technology for integration into UDT's bioterrorism-detection device, and thus, significantly accelerated the development and deployment of UDT's bacterial spore monitor, the BSM-2000.

Ponce saw great potential for this commercialization of NASA technology. He commented, "I see this technology being used in any large public space, possibly working in concert with other detection technologies to provide complete coverage of potential attacks with pathogens."

Product Outcome

Since UDT has the exclusive option on all intellectual property related to the deveopment of the ASD, it has made great strides to get the BSM-2000 on the market.

The unit operates on fairly simple principles. It combines a bioaerosol capture device with a simple analysis method, a robust chemical test for bacterial spores. Airborne particles are drawn to the machine through a fan, captured, and then tested for weaponized anthrax. In addition to its simplicity, the unit has a host of other benefits. It is a long-term solution that runs continuously with little maintenance. It requires very little in the way of operating costs, and has a high reliability factor, with low susceptibility to false alarms.

It is portable. The average unit is just about the size of a computer terminal and weighs less than 50 pounds. The unit consists of an air sampler, a continuous glass fiber tape for capturing spores, a motorized heating element to lyse spores, a syringe pump to deliver the reagent, and a spectrometer to detect luminescence.

UDT has already sold these units around the world. Sartec Saras Technologie is its exclusive Italian distributor. Quantum Automation handles distribution for Singapore, Malaysia, and Thailand. On the home front, Securewrap is making plans to install the monitor at its facility at the Miami International Airport, one of the busiest in the country and a major hub of international cargo shipments.

UDT is working with major hotels, Las Vegas casinos, and convention centers. The company is also in discussions with the U.S. State Department to get the BSM-2000 into mail rooms at embassies overseas and has presented the technology to the United States Postal Service. The unit also has great potential for use in train stations, shopping malls, sports arenas, and any place where large groups of people gather.

Soon, the units will be spread all over the world and working as an effective first step in the detection of biological attacks. If the monitor is activated, and there is an anthrax attack, authorities will know within minutes, not days, and lives will doubtless be saved. Authorities can prevent the sicknesses, deaths, and disasters, rather than having to respond to them. ❖

Lightning Often Strikes Twice

Originating Technology/ NASA Contribution

Contrary to popular misconception, lightning often strikes the same place twice. Certain conditions are just ripe for a bolt of electricity to come zapping down; and a lightning strike is powerful enough to do a lot of damage wherever it hits. NASA created the Accurate Location of Lightning Strikes technology to determine the ground strike point of lightning and prevent electrical damage in the immediate vicinity of the Space Shuttle launch pads at Kennedy Space Center.

The area surrounding the launch pads is enmeshed in a network of electrical wires and components, and electronic equipment is highly susceptible to lightning strike damage. The accurate knowledge of the striking point is important so that crews can determine which equipment or system needs to be retested following a strike. Accurate to within a few yards, this technology can locate a lightning strike in the perimeter of the launch pad. As an added bonus, the engineers, then knowing where the lightning struck, can adjust the variables that may be attracting the lightning, to create a zone that will be less susceptible to future strikes.

Partnership

In 2000, Consumer Lightning Technologies, Inc., of Asheville, North Carolina, licensed from NASA the lightning detection technology that was developed at Kennedy. It calls this project the Precision Lightning Strike Location System, and like many good scientific endeavors, it has an abbreviation: PLSLS. Consumer Lightning Technologies was later absorbed by Mag Holdings, Inc., which created the company, Zap Guard USA, Inc. A lightning products and research company, Zap Guard USA is continuing the research and now marketing the PLSLS.

Sam Gasque, past president of Consumer Lightning Technologies and a technical consultant for Zap Guard USA, is still active in the research and marketing phases of the PLSLS. He believes this technology has strong potential for a variety of venues, virtually any place that is prone to lightning strikes. He sees great possibilities for the use of PLSLS at utility companies, airports, government agencies, golf courses, industrial complexes, and racetracks; on farms and in small towns; and as a claims- verification tool for insurance companies.

With the Precision Lightning Strike Location System (PLSLS), crews can identify and eliminate the source of lightning attraction.

Product Outcome

Zap Guard USA wants to know exactly where lightning strikes, so it can determine what is attracting the lightning. Its research is focused on airports and, specifically, airport runway lighting systems.

Larger airports may have thousands of miles of wire and cable underground, and those that were formerly U.S. military air bases could have 10 times more wire and cable underground than needed. This old, abandoned cable could easily draw a lightning strike and provide a path for lightning to spread to other places within a facility and destroy all electronics in its path. The PLSLS could show maintenance crews the location where lightning has hit. Crews could then correct the situation, usually by digging up unused cables and pipes, so that the attraction to lightning is minimized in the future.

Gasque, back in the 1980s, also developed the Lightning Retardant Cable (LRC) with the help of Kennedy and the Southern Technology Applications Center. This patented technology helps protect electrical wires and cables from strikes, and was featured in *Spinoff* 1998. It keeps lightning from traveling through the cable, preventing damage to satellites, antennas, and cable systems. The LRC is approved by the U.S. Federal Aviation Administration and recommended by one of the largest insurance companies in the United States.

Zap Guard USA plans to offer the PLSLS technology to airports as a companion product to the LRC. From there, it will expand its campaign to power utilities, sports facilities, defense, industry, and government.

Gasque believes that one of the major social benefits of the new PLSLS technology is verification of insurance claims. "Each year, claims due to lightning strikes continue to climb. Where insurance companies once would simply pay a claim, today they tend to verify the loss first, before settlement. Individuals may now need to prove there was a lightning strike in their area before they can recoup damages," Gasque explained. "Insurance companies will call the weather service or a similar service to find out if there was a thunderstorm in a claimant's area on the specific day of loss. If not, the claimant may have trouble getting paid for the damages," Gasque added. As PLSLS is implemented in given areas, exact information regarding lightning strikes can be given to either the consumer or the insurance company.

According to Gasque, once PLSLS is implemented further, sporting events will become safer, as will air travel and traffic signals. Utilities will also become more reliable during storms. All of these improvements will have a positive economic impact.

The engineers at Zap Guard USA have been pushing the technology for the past few years to make it simpler and more effective. They have, with the help of electrical engineering students at North Carolina State University under the guidance of Dr. Bart Greene, been making major advances to the technology. When they first licensed the device, it was hard wired and had a range of less than 1 mile. Now, a few years later, it is wireless, solar powered, and has a range of more than 30 miles.

As Gasque points out, "The bottom line of PLSLS is once you know exactly where lightning strikes, you can identify the attraction. After you remove or neutralize the attraction, the frequency of strikes is substantially minimized. Minimize the strikes and you create a safer social and industrial environment with minimal loss." ❖

NASA created this technology to determine the strike location of lightning around the launch pads so that crews could test the wires and electrical components after a storm.

Technology That's Ready and Able to Inspect Those Cables

Originating Technology/ NASA Contribution

Attempting to locate a malfunctioning wire in a complex bundle of wires or in a cable that is concealed behind a wall is as difficult as trying to find a needle in a haystack. The result of such an effort can also be costly, time-consuming, and frustrating, whether it is the tedious process of examining cable connections for the Space Shuttle or troubleshooting a cable television hookup. Furthermore, other maintenance restrictions can compound the effort required to locate and repair a particular wiring problem. For example, on the Space Shuttle, once a repair is completed, all systems that have a wire passing through any of the connectors that were disconnected during troubleshooting are affected and, therefore, must undergo retesting—an arduous task that is completely unrelated to the original problem.

In an effort to streamline wire inspection and maintenance, two contractors supporting NASA's Kennedy Space Center invented the Standing Wave Reflectometer (SWR) in 1999. In doing so, they leveraged technology that was first developed to detect problems that could lead to aircraft accidents, such as the one that resulted in the catastrophic failure of TWA flight 800 in 1996. The SWR performs a non-intrusive inspection that verifies the condition of electrical power and signal-distribution systems inside the Space Shuttle orbiters. Such testing reduces processing delays and ensures safe operation of these systems.

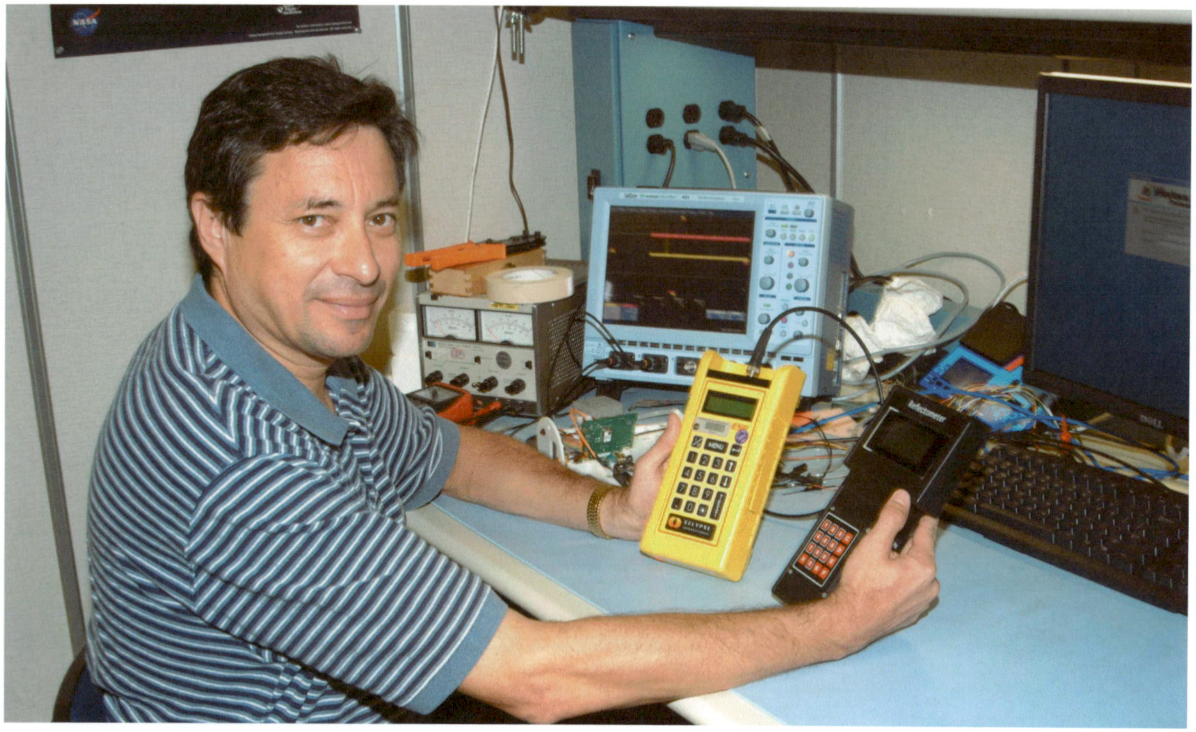

Pedro Medelius, chief technologist for ASRC Aerospace Corp., helped invent the Standing Wave Reflectometer (in right hand). The commercial version of this device (in left hand) was spun off by Eclypse International Corp.

Commercial non-intrusive cable testers that preceded the SWR proved ineffective for wire examination on the Space Shuttle, because of the high-frequency test signals they produced. These signals were incapable of penetrating the dedicated signal conditioners (DSCs) that are used on the Space Shuttle to modify transducer outputs and other signals to make them compatible with orbiter telemetry, displays, and data-processing systems. Recognizing that the need for a low-frequency test was critical to the success of the Space Shuttle, the inventors of the SWR ensured NASA that their hand-held technology could easily pass through DSCs without any interference from signals outside of the intended bandwidth.

When troubleshooting potential instrumentation problems in the orbiter, NASA technicians use the SWR tester to find the precise location of a short or open circuit in a cable. This ability saves NASA many hours of testing on Space Shuttle by minimizing repair, retest time, and any future troubleshooting efforts.

Eclypse International Corp., a leading developer of automated test equipment and associated test application software solutions for commercial and military organizations worldwide, further developed the SWR for use outside of NASA. The resulting technology is currently being used to test wires and cables in complex vehicle systems, and has the potential for myriad consumer-driven applications, as it features a menu-driven interface simple enough for a novice to use.

Partnership

Kennedy's Technology Transfer Office partnered with the Research Triangle Institute and the Southeast Regional Technology Transfer Center at the University of Florida to market the patented SWR technology. The Technology Transfer Office held a briefing at Kennedy, where the SWR technology was demonstrated to more than 50 manufacturers of electronic test equipment, including Corona, California-based Eclypse. Following the briefing, each of the participating manufacturers submitted a license application that included a developmental plan and a market-application plan. Eclypse, which had been exploring useful hand-held troubleshooting tools that could augment and assist maintenance personnel in locating short- and open-circuited wiring in aviation vehicles, was awarded the exclusive patent rights to the NASA SWR technology. Eclypse felt that the SWR would provide a reasonable, low-cost design concept with capabilities not found in the higher-priced cable reflectometers that were currently available.

Two key factors contributed to Eclypse's successful procurement of the exclusive NASA license. First was an open-architecture design concept that would allow for the integration of future technologies into the device, and second was a requirement that would provide the company with access to the principal inventor of the SWR during the beginning stages of product development. Eclypse believed that involvement of key personnel during development of its ESP commercial fault location analyzer was vital for establishing an effective businessplan for the product.

Product Outcome

Although the ESP tag for Eclypse's commercial handheld tester does not stand for "extrasensory perception," the product comes pretty close to having a "sixth sense" in the way that it detects electrical malfunctions with speed and precision. The company claims that the ESP is the fastest and easiest-to-use hand-held wire tester available today.

The Eclypse hand-held, non-intrusive analyzer finds the precise location of a short or open circuit in a cable.

Eclypse has developed three configurations of its ESP product line based upon NASA's SWR technology. The original model, the ESP, provides a basic tool to locate hard faults, such as opens and shorts, and has proven to be very effective for reactive maintenance events. This tool is preferred by commercial airlines. The second-generation model, ESP+, retains all of the features of the original, but also contains a flash memory that allows the operator to store 32 test waveforms that can be uploaded to any personal computer for comparison and analysis. The ESP+ is preferred by military and aerospace organizations for its data-collection capability. The third configuration, the ESP Scientific model, is a 100 MHz, +/- 250VDC version of the ESP. This model provides a loading capability that has proven to be very effective in detecting "soft" faults, such as corrosion and degradation defects that can happen prior to a wire breaking.

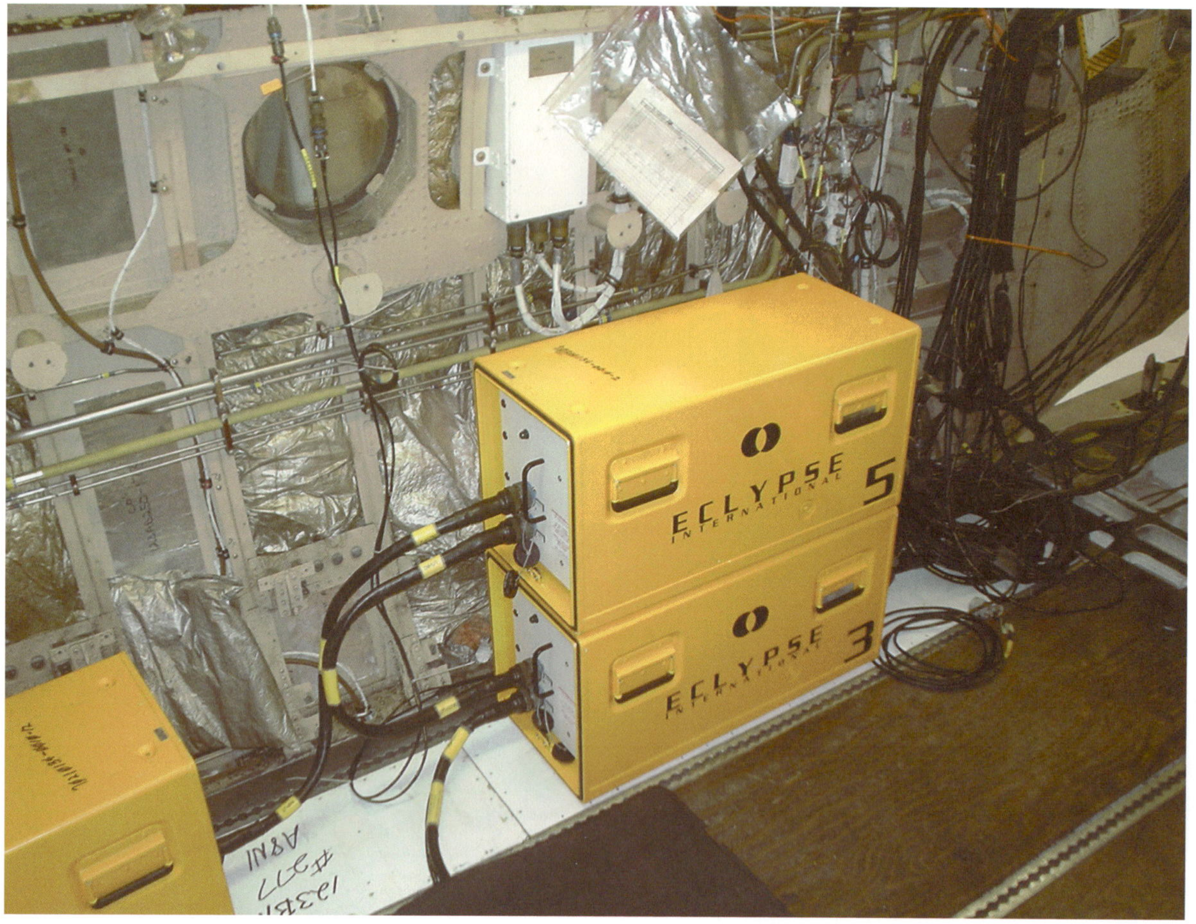

The *C-2 Greyhound* full aircraft circuit-analysis test at North Island Naval Air Depot, Coronado, California.

The ESP family of hand-held SWR meters provides the means to determine the nature and location of discontinuity in a cable by monitoring the impedance while injecting an oscillating signal at one end of a cable and varying the frequency of the oscillating signal. A numerically controlled oscillator generates the signal, and a microprocessor automatically makes the desired measurements and displays the results.

Eclypse has made many improvements to the original NASA prototype hand-held device. The NASA prototype was physically large, fragile, and required specific hexadecimal keypad entries to perform limited functions. In contrast, the ESP weighs only 1 pound, 3 ounces, and measures a mere 4 inches wide, 8 inches high, and 1 inch deep. (NASA has since replaced its hand-held tester with the ESP for Space Shuttle maintenance and test requirements.) Additional features integrated by Eclypse include an alphanumeric, selectively back-lighted liquid crystal display and keypad, a rechargeable battery, automatic shutoff power management, a serial data port, and a drip/splash-proof and shock-resistant case.

The ESP has a specified range of 1,000 feet and can operate successfully in temperatures ranging from -20 to +60 °C. It also provides the capability to store 99 programmable settings for various cable and conductor types, including coaxial, triaxial, multistrand single conductor, and twisted pair. These presets can be recalled quickly for testing, using the simple menu-driven interface. Best of all, the ESP presents solutions in plain English language and does not require visual waveform analysis to determine the cause and location of a discontinuity.

The ESP meter will benefit any industry that generates, distributes, or consumes electrical energy to control processes. Commercial customers include Emerson Network Power, a manufacturer of precision-controlled, uninterruptible power systems that include air conditioning and fluid cooling systems, and APW Ltd., a designer, manufacturer, and integrator of electronic products. In aerospace and aviation, Boeing Aircraft, Lockheed Martin Corporation, Delta Airlines, and Continental Airlines have applied the technology.

At Seymour Johnson Air Force Base in North Carolina, the 4th Component Maintenance Squadron is writing software programs for the ESP tester to troubleshoot aircraft wiring within 3 minutes, as opposed to the 20-plus hours sometimes needed to manually complete such a task with a conventional multimeter.

"The Eclypse tester is more accurate, faster, and easier to use than anything I've seen, and we simply can't build anything that can do what it can in this shop," claimed Master Sergeant Stephen Hoggard, 4th Component Maintenance Squadron chief, electronic and environmental section.

In another military application, the ESP+ model is ensuring the safety of U.S. troops serving overseas. In October 2004, a U.S. *EA-6B Prowler* aircraft deployed in Bagram, Afghanistan, required extra attention, because of intermittent navigational problems within a critical cockpit display panel. These problems occurred multiple times on nearly every flight, though were rarely observed on deck during routine troubleshooting exercises. Most importantly, they compromised the safety of flight, since the plane was flying in a war zone, through mountainous terrain at night.

The U.S. squadron maintaining the *Prowler* aircraft troubleshot the problem for weeks, changing all associated removable components and performing wire checks, all without success. To add to the difficulty, documentation detailing the circuits for the problematic display panel was not available.

To facilitate further troubleshooting, the squadron decided to remove the wing skin panels and expose the wire bundle containing the suspect circuits, so that the bundle could be physically manipulated while under electrical test.

The squadron tested the exposed wires with a common digital multimeter, a time domain reflectometer, and then a hipot insulation-resistance tester, but could not determine the "culprit" wire. Next, it performed a test with the ESP+ meter, and the faulty wire that was the root of the navigational problems was revealed. The wire was replaced and the problem was solved. The success with the ESP+ motivated the squadron to use the meter to identify several other long-standing discrepancies on other aircraft.

The ESP Scientific model is currently used by the Federal Aviation Administration for impedance spectroscopy (measuring corrosion rates) and is preferred by formidable research teams that include U.S. military branches, NASA, and aerospace firms.

All three ESP products are assuaging White House Office of Science and Technology Policy (OSTP) concerns about aging wiring. The OSTP concludes that such issues extend beyond aviation and are of national concern, while referring to the SWR-spinoff technologies as "intelligent systems for identifying faults."

On the whole, the economic benefits derived from the application of SWR technology include: lower investment costs of troubleshooting equipment; reduced time and effort to affect repairs; reduced time to validate repairs; and avoidance of failure-induced downtime, all of which result in lower total operating cost. ❖

Delta Airline technicians Dane Swenson and Jim Elick use the Eclypse analyzer to test generator feeders on a Boeing 767 commercial aircraft.

Secure Networks for First Responders and Special Forces

Originating Technology/ NASA Contribution

When NASA needed help better securing its communications with orbiting satellites, the Agency called on Western DataCom Co., Inc., to help develop a prototype Internet Protocol (IP) router.

Westlake, Ohio-based Western DataCom designs, develops, and manufactures hardware that secures voice, video, and data transmissions over any IP-based network. The technology that it jointly developed with NASA is now serving as a communications solution in military and first-response situations.

Partnership

In early 2000, Glenn Research Center approached Western DataCom to develop the prototype IP router. This was part of NASA's "IP in Space" initiative, which looked to employ commercial off-the-shelf products to support reliable, fast, and secure communications between NASA and its orbiting satellites. The company signed a Space Act Agreement with Glenn and delivered a prototype device that met the three requirements set by the NASA research center, namely speed, security, and reliability. The router employed advanced data-compression techniques (to improve throughput and meet the speed requirement) and encryption (to meet the security requirement), and operated with commercial protocols (to meet the reliability requirement).

Because of the work it had done for Glenn, Western DataCom was approached by Cisco Systems, Inc., in 2001, to participate in the development of an IP encryptor for the Cisco Mobile Access Router (3200 Series), for military use. According to Western DataCom, it offered Cisco two distinct advantages: 1) Western DataCom had leading encryption and compression technologies, from working with NASA, as well as the National Security Agency; and 2) Western DataCom code developers possessed the military clearances needed to perform the work required. Cisco created its "Advance Technology Partner" classification and named Western DataCom the first of such partners. Cisco also joined Western DataCom in working with Glenn to develop the reliable, fast, and secure mobile router system for military and first-response use.

The technology was not commercially available at the time that the September 11 attacks took place, but will prove invaluable to emergency and rescue personnel in averting any potential future threats.

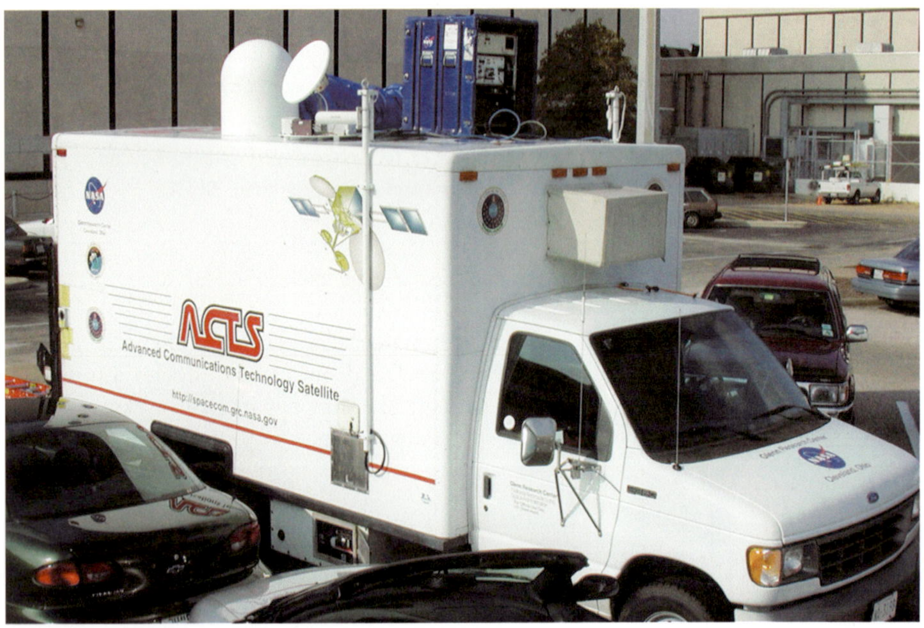

This van served as the mobile test bed when the IP router technology was field-tested at Glenn Research Center.

Product Outcome

In the hours and days after September 11, communications between first responders and emergency-management officials from Federal, state, and local agencies were severely disrupted. New York City's Emergency Operations Center, designed to coordinate rescue efforts in a major terrorist attack, was housed in the 47-story "7 World Trade Center" building and destroyed.

The World Trade Center was a node of central communications for all forms of voice and data traffic and was utilized by business and private customers, as well as the city's first responder and emergency-management agencies. Communications systems for the police and fire departments were temporarily disabled as a result of the damage caused by the collapsing of the building and senior emergency-management officials were unable to contact first responders in the early hours of the tragedy.

Because police and firefighters could not communicate directly with each other, many firefighters within striking distance of safety never received a police warning on the impending collapse of the South Tower. A report from the University of New Hampshire* concluded that this lack of interoperability between the police and fire communications systems were "at least, partially responsible for the loss of 343 firefighters at the World Trade Center."

Much of New York City's landline and cell phone infrastructure was also damaged or destroyed during the attacks. Moreover, the disaster generated so much communications traffic in and around the city that the remaining intact landline, cellular, and two-way pager systems became too congested to be of use to first responders and emergency-management personnel.

The experiences of September 11 have driven many organizations and individuals to realize that new communications systems are needed to secure our country and improve our ability to respond to terrorist attacks. In addition, the ongoing conflicts in Afghanistan and Iraq have broadened the need to provide a mobile, interoperable, and secure communications system solution for the U.S. military and first responder personnel, such as U.S. Army National Guard, firefighters, police, and emergency medical services (EMS).

In 2004, the secure mobile router system co-developed by Western DataCom and NASA was successfully used by the Army for an aerostat (balloon)-based radar, called the Persistent Threat Detection System, in Operation Iraqi Freedom. The system permits military technical operations centers (TOCs) in Iraq to send secure, high-speed voice, video, and data communications to the field through tactically deployed mobile units. This was the first use, during war, of technology enabling TOCs and mobile units to send secure voice, video, and data communications, according to Western DataCom.

First responders from Cook County, Illinois; the New York Port Authority; and the New Jersey Port Authority are currently utilizing the company's secure system in preparation for natural or man-made disasters. Also, in 2004, Western DataCom developed a secure-communications modem to be utilized primarily by first responders for homeland defense operations. These products were successfully deployed during the Republican National Convention and the presidential inauguration.

Western DataCom Co., Inc.'s Executive Travel Case sets up connections automatically to the Internet, Secret Internet Protocol Router Networks (SIPRNETs), and Non-secure Internet Protocol Router Networks (NIPRNETs), for use by the U.S. Joint Forces Command.

Recently, Western DataCom received a $100,000 Glenn Alliance for Technology Exchange (GATE) award from Glenn and Battelle, an organization that helps bring NASA technology to companies outside the traditional aerospace industry. The award, in the form of $50,000 in cash and $50,000 in Glenn engineering time, will be used to design a small personal computer encryptor card for commercial markets. This card is anticipated to act as a shield outside of a computer, protecting its hard drive from outside "attacks," such as worms and viruses, as well as "middle-man" and "spoofing" threats. (A "middle-man" is someone who unwittingly spreads a virus by simply opening or forwarding an e-mail, while "spoofing" is a technique used to gain unauthorized access to computers. A user receives e-mail that appears to have originated from one source when it actually was sent from another source, in an attempt to trick the user into releasing sensitive information.)

The two NASA engineers assigned to this project have experience with Western DataCom, in that they were involved with the 2000-2001 "space router" project that culminated in the basic technology platform utilized in the company's current encryptor product line.

Once the commercial personal computer card is operational, Western DataCom plans to design a top-secret military version. The company intends to have the commercial card designed and operating, and to have work started on the military version, by the end of 2005. ❖

*Lund, Donald A., The Lessons of Non-Interoperability in Public Safety Communication Systems. University of New Hampshire, April 2002.

Space Suit Spins

Originating Technology/ NASA Contribution

Space is a hostile environment where astronauts combat extreme temperatures, dangerous radiation, and a near-breathless vacuum. Life support in these unforgiving circumstances is crucial and complex, and failure is not an option for the devices meant to keep astronauts safe in an environment that presents constant opposition. A space suit must meet stringent requirements for life support. The suit has to be made of durable material to withstand the impact of space debris and protect against radiation. It must provide essential oxygen, pressure, heating, and cooling while retaining mobility and dexterity. It is not a simple article of clothing but rather a complex modern armor that the space explorers must don if they are to continue exploring the heavens.

The XMC-2 ILC Full Pressure Suit (1955-1959) was built under a contract with the U.S. Air Force and incorporated bellows and convolutes that greatly increased mobility. This suit laid the groundwork for the design of the Apollo space suits.

Partnership

In 1947, before people successfully left the Earth's atmosphere, ILC Dover, Inc., of Frederica, Delaware, the governmental supply branch of the International Latex Corporation, was manufacturing high-strength rubber rafts for U.S. Navy Seals. This product proved the basis for a succession of increasingly complex projects in the high-strength textile market. In the next decade, the company's production of pressurized textiles brought it into the realm of creating pressurized flight suits and helmets for American U-2 spy plane pilots, who flew in extreme altitudes. ILC Dover's success with pressurized flight suits led to its bidding and winning the space suit contract for the Apollo mission in 1965 with its AX5L suit—a modified, pressurized suit with woven restraints, air bladders, and wrist cuffs that still is used by astronauts today.

Since the early Apollo contract, ILC Dover, in conjunction with Hamilton Standard, of Windsor Locks, Connecticut, has designed and produced space suits for NASA. These suits have flown on every mission, and ILC Dover maintains a staff of 15 full-time employees at the Johnson Space Center dedicated to fitting, assembling, and repairing the space suits for the entire astronaut corps.

ILC Dover's partnership with NASA to create the space suits resulted in mastery of life-critical applications, quality workmanship, and testing for high reliability and safety in extreme conditions, as well as the repeated analysis of processes and inspection. ILC used these skills and the NASA-honed expertise to create a handful of spinoffs, including work in the medical, pharmaceutical, personal protection, and aerospace markets.

NASA also called upon ILC Dover's skills recently, providing the company's engineers the opportunity to spin some of their technical talents back into the Space

In 2004, ILC Dover, Inc.'s new I-Suit, an extra vehicular activity space suit, was tested in Meteor Crater, Arizona.

Program. They designed and manufactured the Mars landing space inflatables for the Pathfinder and Mars Explorer Rover (MER) Missions. These large airbags cushioned the drop onto the surface of the Red Planet. In addition, they manufactured the landing ramps and continue to provide suits to support the construction of the International Space Station.

The I-Suit is a highly mobile, soft extra vehicular activity space suit that accommodates astronauts from the 5th percentile female to the 95th percentile male.

Product Outcome

Through its work with space suit designing, testing, and manufacturing, ILC Dover developed skills and processes unique to industry. It has harnessed this space know-how on numerous occasions and has developed several spinoff products. The work ranges from therapeutic suits to safe, efficient pharmaceutical manufacturing; to new, simple-to-use, life-saving gas and chemical masks; and to lighter-than-air (LTA) vehicles.

ILC Dover has engineered therapeutic cooling and heating suits to assist people suffering from various maladies. The Cool Vest, featured in *Spinoff* 1982, is a lightweight cooling garment, designed to reduce the effects of heat stress, that ILC Dover created out of its work with the liquid-cooling systems used to regulate astronaut body temperature in the space suit. This vest has applications in rigorous industrial environments where elevated temperatures can be debilitating. It also has been used to lower the body temperatures of people suffering from hypohidrotic ectodermal dysplasia, a rare disease in which the sufferer is lacking in sweat glands, as well as to lessen the effects of heat-related symptoms in people suffering from multiple sclerosis.

More recently, the engineers at ILC Dover experimented with using the same radio frequency molecular stimulation that they use to fasten space suit components together to generate low levels of heat deep within muscles for treating muscular stress. The stimulation caused by this technique leads to deep tissue diathermy or the oscillating of cellular muscle tissue for deep warmth. Whereas a conventional heating pad applied to an injury would only heat the surface of the ailment, the radio frequency approach manages to heat the actual sore tissue. Developed in conjunction with SeliCor, Inc., a biomedical device company in Austin, Texas, this product is now being marketed as SeliTherm.

ILC Dover has pushed applications of its space know-how into the pharmaceutical-manufacturing arena with the creation of the DoverPac. Designed in collaboration with pharmaceutical giant, Eli Lilly and Company, the DoverPac provides high levels of reliable containment during charging and offloading of highly active ingredients in powder form. The system, a series of high-strength, flexible, transparent tubes, helps get drugs to the market faster while protecting workers and ensuring integrity with its low permeability and many safety features. It is composed of a tough fabric outer restraint and a durable liner made of the rugged ILC Dover-developed ArmorFlex film. At the top of the many safety features, this system offers increased productivity with a uniquely designed crimping system that allows one-person operation and integral sleeves that facilitate quick sampling without breaking containment.

A great deal of NASA knowledge went into the production of this product with the potential of revolutionizing

Mars lander with ILC Dover, Inc.'s deflated airbags that cushioned the rover's landing.

ILC Dover, Inc. has provided over 1.6 million M40 gas masks to the U.S. Army.

the pharmaceutical-manufacturing industry. First, the DoverPac is manufactured at the same factory, and undergoes the same scrutiny and rigorous testing as NASA's space suits. Furthermore, the heat-sealing technology employed by ILC Dover to seal components within the space suit is the same that is used to create the DoverPac, and the woven restraint/bladder that contains the powderized product mirrors the design approach used on the space suit. The outer restraint of the DoverPac is made of a grounded woven mesh to avoid static sparks near the fine powders. This is the same mesh found on the inflatable devices used for the Mars Pathfinder and MER.

ILC Dover takes great pride in knowing that it provides state-of-the-art protection from chemical, biological and nuclear inhalants for U.S. soldiers. Already the sole provider of the M40 gas mask to the U.S. Army and having equipped more than 1.5 million soldiers to date, ILC Dover has developed a new civilian gas mask, the SCape Hood.

M40 masks are the best protection available for trained soldiers, but are not practical for the average person. Traditional gas masks employ face seals that will not work on bearded or unshaven people or wearers of eyeglasses. Moreover, traditional gas masks demand multiple sizes to fit the wearer population, require extensive training, and are not designed for children or infants. The SCape Hood, however, eliminates these problems. Due to its unique design, there is no worry about size or seal, and the device can be in place and functioning in under 30 seconds. As soon as the mask is removed from its container, a blower automatically initiates and the hood inflates with clean air and creates sealing and positive pressure. Because of the blower, the unit also can be used on infants and the unconscious. The blower will operate for 4 hours.

The same heat seals used for the harsh conditions of space bind the SCape Hood together. More important is that the same quality and inspection trusted to keep the astronauts alive is employed to create this life-critical product. The expertise gained from protecting the astronauts from the harsh space environment transfers directly to work in protecting people on Earth from chemical warfare and hazardous material spills. Currently, ILC Dover has sold over 60,000 SCape Hood units. The company anticipates that the SCape Hood will be in high demand in urban areas and in government buildings possibly targeted by terrorists.

ILC Dover's use of high-performance, rugged textiles has led it in yet another direction. The engineers have been using their space experience to manufacture blimps, aerostats, airships, and other LTA vehicles since the early

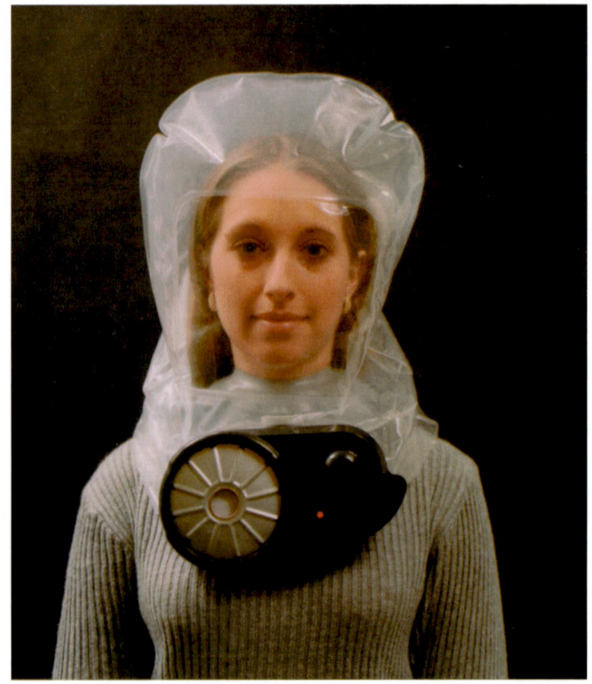

The SCape Hood can be fitted in place and functioning in 30 seconds or less.

1970s. In fact, currently 80 percent of the large LTA structures in the world are manufactured by ILC Dover at the same location as the space suits. These structures employ high-strength seaming technology designed and developed for the suits. Test methods developed for the films used on the space suits are the same methods employed to test the materials used on the structures. The high-strength, ultraviolet-resistant, low-gas permeability technology necessary to float these crafts at high altitudes and not lose gas or suffer damage from ultraviolet radiation is a direct result of experiments done to determine the types of textiles and engineered films best suited for the harsh environments of the space landscape.

The LTA vehicles vary from the typical blimp often seen hovering above sporting events to the TowTech

Lifting Sphere, a massive inflatable device that harnesses the lifting power of helium for use in areas where roads or rivers are not accessible. These LTA structures range in volume up to 4 million cubic feet and include tethered surveillance aerostats, advertising airships and blimps, passenger airships, high-altitude airships, and hybrid heavy-lifting vehicles.

ILC Dover has partnered with NASA from the very beginning and continues to support its missions. NASA recently called upon ILC Dover for its expertise in the manufacture of extreme-situation textiles, and the company responded by designing and manufacturing inflatable airbags.

ILC Dover, Inc. manufactures 80 percent of the LTA in the world. Blimps, like the one pictured here, often fly over sporting events and vacation spots.

In addition to developing myriad spinoffs that came from the space suit work, the continued support of NASA's astronaut corps with flight suits, and the work related to the exploration of Mars, ILC Dover is designing another space suit. This new suit, the Extravehicular Mobility Unit (EMU), is lighter and offers a greater range of motion than the older suits.

Safety and quality are always a focus at ILC Dover. As an ISO 9001:2000-registered company, ILC Dover brings more than half a century of experience to the design, development, and manufacture of life-critical products. From space exploration to physical therapy, from defense to pharmaceutical, ILC Dover develops real-world solutions for a vast range of industries.

The challenges of the future are what compel ILC Dover forward. Man landed on the Moon—ILC Dover created the space suit. Researchers develop wonder drugs—ILC Dover aids in safe production. Rovers explore Mars—ILC Dover helps them land safely. With an enviable record of performance, a team of highly qualified personnel, and a sound foundation of proven and innovative technologies, ILC Dover continues creating what's next through spin-ins and spinoffs. ❖

Cool Vest™ is a trademark of ILC Dover, Inc.
SeliTherm™ is a trademark of SeliCor, Inc.
DoverPac® is a registered trademark of ILC Dover, Inc.
ArmorFlex® is a registered trademark of ILC Dover, Inc.
SCape Hood™ is a trademark of ILC Dover, Inc.

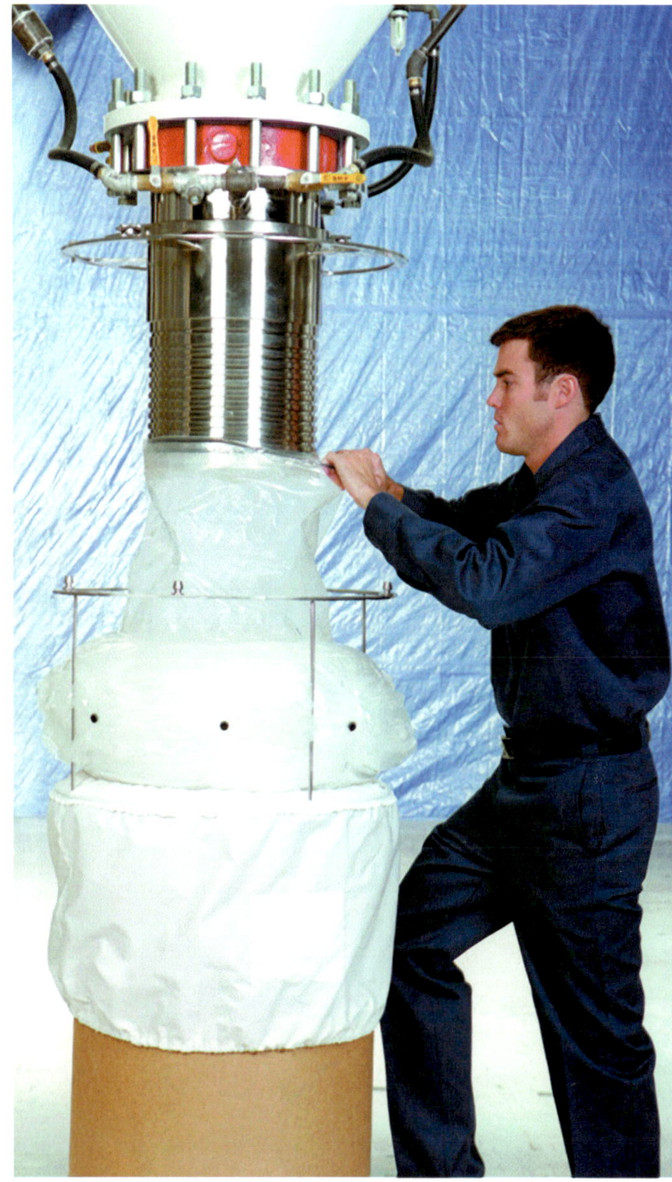

A technician works with the DoverPac system, a revolutionary, safe pharmaceutical-manufacturing system.

Cooking Dinner at Home—From the Office

Originating Technology/ NASA Contribution

It is well past quitting time, but you are still stuck in the office. Your spouse left work over an hour ago, but is caught in bumper-to-bumper traffic. As a result, neither of you were available to pick up your daughter on time from her soccer game. If your son hadn't gotten detention at school today—which also made him late for work—he could have picked her up.

The next thing you know, it is already 8:30 at night, and your family members are finally all together under the same roof. No one has had a bite to eat since lunch, and dinner certainly isn't going to cook itself…or is it?

For those who are all too familiar with this situation, it might be time to welcome the oven of the future into your homes: the *Connect*Io Intelligent Oven, brought to you by TMIO, LLC, of Cleveland. Applying the same remote command and control concepts that NASA uses to run experiments on the International Space Station (ISS), *Connect*Io allows its owners to cook dinner from the road, via a cell phone, personal digital assistant, or Internet connection.

By combining remote-access technology with the capability to both cook and refrigerate food, TMIO, LLC, has eliminated the "wait time" of cooking.

Partnership

In 1994, David Mansbery was at the helm of an active family whose schedule rarely allowed for home-cooked meals. Growing tired of frequent fast-food dinners, Mansbery set out to bring traditional, home-cooked dinners back into his home. At the time, Mansbery was president of a natural gas supply company, and NASA's Glenn Research Center was one of his biggest clients.

When Mansbery pitched his idea of a hot-and-cold, remotely operated oven to Glenn, the NASA center lined him up with a group of its engineers that had worked on the ISS Electric Power System, the Solar and Heliospheric Observatory (SOHO) that studied the Sun, the Cassini spacecraft that is currently visiting Saturn, and several experiments that flew on the Space Shuttle. These engineers supplied Mansbery with "Embedded Web Technology" software that was developed at Glenn in 1996. According to its inventors, Embedded Web Technology marries embedded systems (hardware or software that forms a component of a larger system and is expected to operate without human intervention) and the World Wide Web, to let a user monitor and/or control a remote device with an embedded system over the Internet, using a convenient, graphical user interface. In contrast to general purpose desktop computers, embedded systems contain processors, software, input sensors and output actuators—all of which are dedicated to the control of a specific device.

NASA counts on Embedded Web Technology to allow astronauts to operate experiments from anywhere on the ISS, using any laptop computer available to them. This way, NASA can operate its experiments without having to install user-interface software on all of the laptop computers for every space-bound instrument. To date, an estimated $150 million has been saved by use of this software to control Space Station payloads.

With authorization to use the Embedded Web Technology, Mansbery formed the TMIO company to execute his idea. The NASA software enabled low-cost, real-time remote control and monitoring of the resulting intelligent oven product.

Product Outcome

With combined cooling and heating capabilities, TMIO's *Connect*Io stainless-steel professional series oven provides convenience and healthy living for today's active lifestyles. Before leaving for work, consumers can place fresh food in *Connect*Io, where it will remain properly refrigerated until a programmable cooking cycle begins; the oven is preprogrammed with a universal cooking menu, which can be customized to the user's preferences. The menu allows the user to simply enter in the dinner time, and the oven automatically switches from refrigeration to the cooking cycle, so that the meal will be ready as the family arrives home for dinner.

Should plans change, the embedded software enables the user to conveniently and instantaneously adjust and control the oven settings while away from home, via a cell phone, personal digital assistant, or the Internet. This includes re-refrigeration, warming for a completed meal, and any temperature or time modifications that need to be made on the fly.

*Connect*Io's unique two-cavity design offers independent cooling and cooking of two separate dishes. For instance, chicken can be placed in the top compartment and potatoes in the bottom. Each dish cools and cooks according to its own instructions.

Built-in Ethernet capabilities allow for a seamless connection of *Connect*Io to the outside world, so additional wiring is not required. The oven also features touch screen controls that integrate into flat panels around the house, allowing for a sophisticated and practical way to see if dinner is ready without having to leave a room, or perhaps miss a part of a favorite television show, to check on the oven. Furthermore, TMIO asserts that *Connect*Io's advanced structural design leads to safer and more energy-efficient cooking and self-cleaning cycles.

The technology was recognized as one of TIME magazine's "Coolest Inventions" for 2003, as Reader's Digest's "Best of America" for 2004, and honored with the 2004 "Best of Innovations" award by the Consumer Electronics Association. It has made numerous television appearances on programs like CNBC's "Power Lunch," "Good Morning America," "The View," and ESPN's "Cold Pizza," as well as A&E Television Network's "At Home with the Brave," a show that offers home-renovation remedies to soldiers returning from Iraq and Afghanistan; TMIO contributed a *Connect*Io oven to a deserving soldier and his family, who were featured on the program. TMIO has also announced its engagement of a Beverly Hills-based product placement firm to help bring *Connect*Io to prime time television and Hollywood movie sets.

Built from NASA technology, one-too-many nights of fast food, and some forward-thinking inspired by the cartoon, "The Jetsons" (TMIO openly admits to this influence), *Connect*Io has revolutionized the cooking industry and brought the family back together for dinner. ❖

From a remote location, the *Connect*Io Intelligent Oven can be programmed to automatically switch from refrigeration to the cooking cycle, so that the meal will be ready as the family arrives home for dinner.

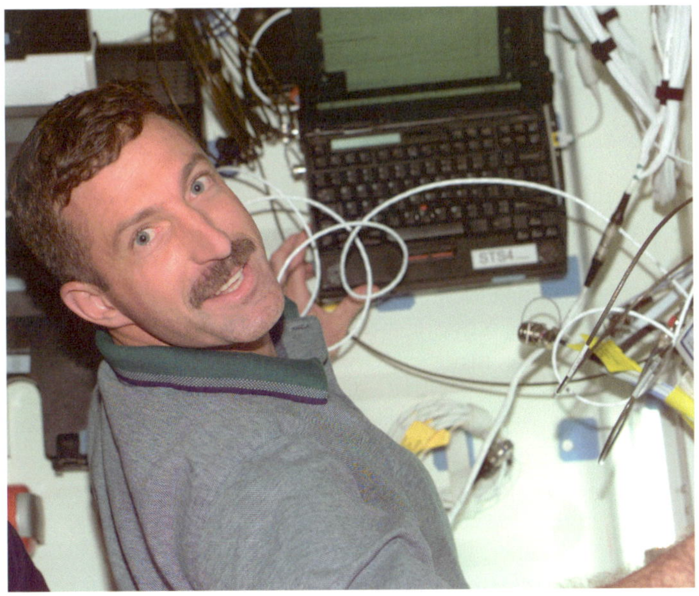

Astronaut Daniel Burbank, mission specialist for STS-106, uses a laptop computer to keep up with busy chores onboard the Space Shuttle Atlantis. Embedded Web Technology, developed at Glenn Research Center, allows astronauts to monitor and operate International Space Station experiments from remote positions, just as it lets owners of the *Connect*Io Intelligent Oven cook meals—away from home.

Nanoscale Materials Make for Large-Scale Applications

Originating Technology/ NASA Contribution

Since its dawning days, NASA has been at the forefront of developing and improving materials for aerospace applications. In particular, NASA requires dramatic advancements in material properties to enhance the performance, robustness, and reliability of its launch vehicles, spacecraft, and the International Space Station. Such advancements over the years include noise-abatement materials, fire-resistant fibers, heat-absorbing insulation, and light-but-strong moldable composites.

In 1991, a new carbon fiber called a carbon nanotube was discovered and fully substantiated by a Japanese electron microscopist. Its dramatic strength and low density (20 times the tensile strength and one-sixth the density of steel) were turning the heads of materials scientists and engineers all around the world, including those who developed equipment for NASA.

While NASA did not invent the carbon nanotube, it is working to advance the fibrous material for widespread, low-cost application in sending humans beyond low-Earth orbit, well into the outer reaches of the universe. Carbon nanotubes have the potential to reduce spacecraft weight by 50 percent or more, by replacing the heavier copper wires currently used, according to NASA scientists. Furthermore, NASA researchers have reported a new method for producing integrated circuits using carbon nanotubes instead of copper for interconnects. This technology has the capability to extend the life of the silicon chip industry by 10 years.

Because of this growing interest in carbon nanotubes and their perpetual possibilities, NASA has funded both internal and external research in this field.

Partnership

In 2001, Zyvex Corporation, of Richardson, Texas, developed a revolutionary method of changing carbon nanotube chemistry without harming the molecular structure that gives the carbon nanotubes their remarkable properties. In February 2003, Johnson Space Center recognized the promise of this technology and began working with Zyvex through a Phase I **Small Business Innovation Research (SBIR)** contract to develop the rational engineering of carbon nanotube surface chemistry.

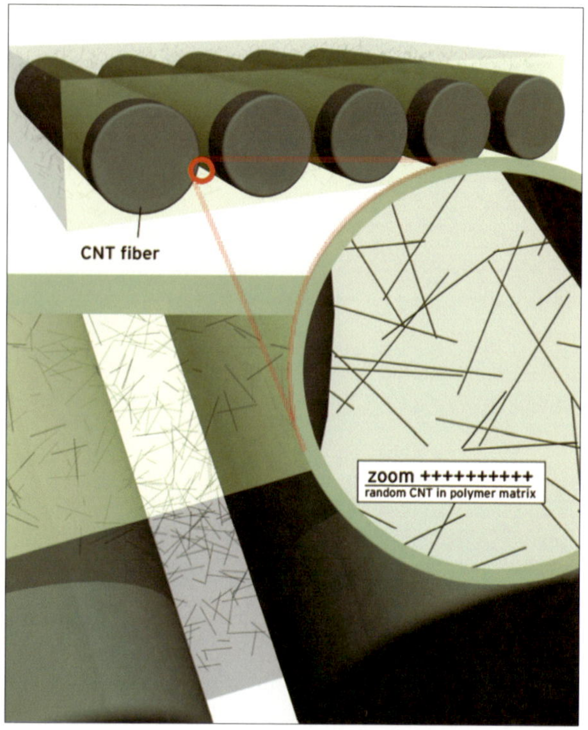

With assistance from NASA, Zyvex Corporation built extremely strong and light hierarchical carbon nanotube-composite materials.

In the Phase I program, Zyvex not only demonstrated the ability to make high-concentration solutions of carbon nanotubes in organic solvents, but also demonstrated the ability to achieve excellent dispersion of the carbon nanotubes in polymers such as epoxy. These successes led to a Phase II SBIR contract with Johnson in January 2004, to build extremely strong and light hierarchical carbon nanotube-composite materials for NASA applications.

Even though the partnership between Zyvex and NASA is still in its early stages, these modified carbon nanotubes are on the fast track for establishing new classes of materials that will meet the Space Agency's needs as it produces new, safer, and more cost-effective spacecraft. The Phase II contract with Johnson is still in progress and a new Phase I SBIR contract with Marshall Space Flight Center to develop carbon nanotube-reinforced radiation shielding is also underway.

Product Outcome

The introduction of a new material into commercial applications is usually a 5- to 10-year process. With the active support of NASA, a rigorous New Product Development Process, and a commitment to speed of execution, Zyvex took less than 3 years to go from invention to commercialization.

Shortly after conceiving and demonstrating a new way to functionalize carbon nanotubes, Zyvex filed initial patents and implemented its New Product Development Process for the carbon nanotube polymer composites.

Based on the successful results of the Phase I Johnson SBIR contract, Zyvex started offering solubilized carbon nanotubes on its Web site in September 2003. Exactly a year later—in the midst of the Phase II SBIR work—Easton Sports, Inc., announced the use of Zyvex carbon nanotube technology in its new line of bicycle parts.

Van Nuys, California-based Easton Sports has an 82-year history of leading its market by developing new materials to achieve enhanced performance in sporting goods. The company's bicycle division has incorporated the Zyvex technology, commercially known as NanoSolve, in its 2005 line of bicycle components, including handlebars.

Robert Folaron, Zyvex's director of product development, noted, "The Easton-Zyvex partnership offers an

Image courtesy of Easton Sports, Inc.

The manipulation of materials on a molecular scale is leading to lighter-but-stronger hockey sticks. Pictured here is Easton Sports, Inc.'s Synergy SL product, featuring Zyvex Corporation's NanoSolve technology.

oppotunity for two leaders to rapidly develop a carbon nanotube-enhanced composite and quickly adopt it into a product line."

John Harrington, vice president of Easton Sports's bicycle division, agreed, saying, "We saw a large increase in the strength and toughness of the composite when the nanotubes were properly functionalized. With that achievement, we were able to create lighter and stronger bicycle components."

Easton Sports again looked to Zyvex and its NanoSolve technology in February of this year, to take hockey and baseball equipment to the next level.

The revolutionary Easton Synergy SL composite hockey stick increases stick strength while reducing weight. It is lighter (420 grams) and stronger than its Synergy predecessor, and has a new blade design that yields "unmatched performance," according to Easton Sports.

"We view the incorporation of carbon nanotubes into our sticks as a critical step in further developing the world's best hockey sticks," said Ned Goldsmith, vice president of Easton Sports's hockey division. "The Synergy SL is the first stick to utilize carbon nanotube's stronger and more durable material design. Keeping with our tradition of putting Easton sticks in the hands of the world's best players, Peter Forsberg has already adopted this new design while playing in Sweden for Modo," he added. Forsberg is also a member of the National Hockey League's Philadelphia Flyers franchise.

Easton Sports is touting its NanoSolve-reinforced Stealth baseball bat as the "best bat in the game today." The NanoSolve technology enhances the company's Opti-Flex composite handle technology, providing maximum handle flex, up to 3 times greater than aluminum bats. Additionally, NanoSolve provides better responsiveness and more "kick" through the hitting zone.

Zyvex recently partnered with several other major corporations to introduce NanoSolve in multiple markets. These businesses include Boston Scientific Corporation; IBM Corporation; Honeywell International, Inc.; Nantero, Inc.; Fiat; and DuPont.

The success realized to date is just the beginning for Zyvex, as it is committed to continuing the development of NanoSolve for both commercial and government applications. Based on the company's preliminary commercial analysis for market segments, it estimates that the potential niche markets for its composites will be modest in the beginning years, but its growth will reach $400 million to $800 million, 10 years after commercialization. Zyvex acknowledges that the market for advanced composites is competitive; but, since its goal is to develop composite materials with strength superior to any material produced today, it believes it will be in a unique position to partner with companies that manufacture carbon and composite fibers.

According to Thomas A. Cellucci, Zyvex president, the company's ongoing relationship with NASA is especially important: "We are extremely pleased to be working with NASA. A large portion of our success in nanomaterials is due to their involvement."

The Federal Aviation Administration will be referencing the Johnson SBIR Phase II contract, according to Zyvex, in a congressional report that highlights the future benefits of carbon nanotubes. Zyvex also has monetary and hands-on support from the Defense Advanced Research Projects Agency to develop high-strength materials. ❖

NanoSolve™ is a trademark of Zyvex Corporation.
Sc900™ is a trademark of Easton Sports, Inc.

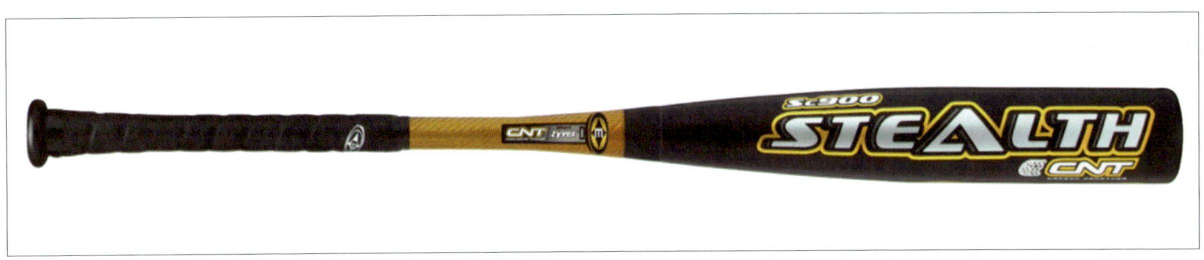

The addition of carbon nanotube technology, made possible by NanoSolve materials and exclusive to Easton Sports, Inc., strengthens composite structures in baseball bats to provide improved handle designs with optimized flex, responsiveness, and more "kick" through the hitting zone, for maximum performance. Pictured here is the Easton Stealth Sc900.

NASA's Growing Commitment: The Space Garden

Originating Technology/ NASA Contribution

Astronauts cannot live on dehydrated ice cream alone. Like everyone else, they need their vegetables. Enter VEGGIE, the Deployable Vegetable System, currently under development by Orbital Technologies Corporation (ORBITEC). VEGGIE is the latest in a long line of vegetable production units ORBITEC is currently working on, with NASA assistance, to grow salad crops to supplement prepackaged foods during long stays in space.

The primary goal of the VEGGIE project is to provide flight crews with palatable, nutritious, and safe sources of fresh food with minimal volume and operational resources. In addition, ORBITEC recognizes the age-old adage that gardening is good for the soul, and it acknowledges that gardens are beneficial for relaxation and recreation. As evidence, astronauts on the International Space Station (ISS), who often stay for periods of 6 months, have been enjoying plant experiments, which provide them with much missed greenery and can occupy valuable free time with an enjoyable task.

VEGGIE is a project that grew out of technology developed by ORBITEC for the Biomass Production System (BPS). The BPS is equivalent in size to a Space Shuttle middeck locker, and provides four plant growth chambers. Each chamber has independent control of temperature, humidity, nutrient and water delivery, lighting, and atmospheric composition.

The BPS flew to the ISS in 2002, and astronaut Dan Bursch had positive comments about his interaction with the plants while in orbit. Astronaut Peggy Whitson had similarly positive remarks during the following expedition while she was growing soybeans for another experiment. Whitson reflects on her time in space with the plantings on Expedition 5, "Although it doesn't sound like much, it was really exciting to see something green. I assumed that this was just because I really enjoy plants, but it surprised me that both of my crewmates were just as excited. They wanted photos of themselves with the plants and asked if they could eat some of them, too!" The astronauts did not eat the plants, but these initial experiments gave the researchers information they needed about the basics of growing crops in space.

Partnership

Since its inception in 1988, Madison, Wisconsin-based ORBITEC has received over 160 government contracts, exceeding a total of $120 million. The VEGGIE work is being done under a NASA Phase II **Small Business Innovation Research (SBIR)** contract, sponsored by Kennedy Space Center. ORBITEC has a long history of NASA involvement, and currently holds the two largest Phase III SBIRs ever issued: one for the Plant Research Unit and the other for the Advanced Animal Habitat, a laboratory for studying rats that is scheduled to fly to the ISS.

Its commitment to NASA and the SBIR program is reflected in the company charter, which outlines the mission to "serve government and industry by developing and demonstrating innovative technologies and advanced products that enhance the quality of human life and support mankind's exploration of the universe."

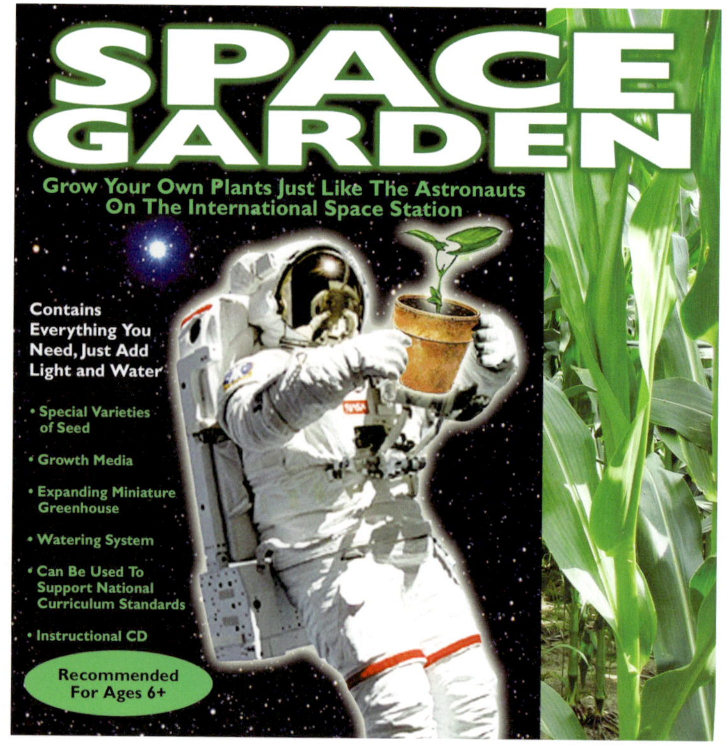

The Space Garden comes with everything students and educators need to begin their own growing experiments at home or in the classroom.

Product Outcome

After the successful BPS flight, discussions began at ORBITEC about innovative methods for growing plants in space to improve plant habitability with extremely reliable yet simple systems. The company had observed astronaut Jim Voss on Expedition 2, the second group to live on the ISS, and astronaut Don Pettit during Expedition 6, the sixth tenancy of the ISS, attempting to grow onions and other varieties of plants in old food bags, but having little success due to the lack of a proper growth medium and fertilizer.

ORBITEC's goal was to produce a device that individual astronauts could take into orbit and use to grow a variety of plants for supplemental food preparation. The VEGGIE and BPS work led to the development of the Astronaut Plant Bag (APB), which seemed the perfect fit, as it uses the same base as the astronaut food containers and, thus, has the same septum holder and hydration needle. The APB can grow food on spacecraft using the existing light sources. In addition, it has a semi-passive atmospheric control system that minimizes water use without severely limiting gas exchange. It is small, uses little resources, and takes virtually no time to maintain. It does not interfere with an astronaut's vigorous work schedule and can be maintained as a hobby; and since APB was built for space travel, it meets all the necessary flight requirements. ORBITEC is expecting to send APBs up with future missions as astronaut personal gear.

The Space Garden spun out of the APB project as an outreach companion and is a method for involving children in biology, life sciences, and agricultural studies. Providing a classroom-based unit that duplicates the projects on the ISS and Space Shuttle, the kit is safe for children ages 6 and up, and is recommended for use in classrooms, grades 5 and up. It is similar to the APB, but less costly, as it does not need to meet the rigorous demands of the flight-tested units.

The Space Garden comes with all of the materials needed to conduct growth experiments, including the growth chamber, seeds, watering syringe, a spongy root barrier, and the growth medium, Lucite. The kit supplies everything an educator needs to make use of this experiment in the classroom, such as an educational CD-ROM, a data log, a classroom presentation, a growth chart, and Web activities. The Space Garden provides teachers with four experiments correlated to National Science Standards. These include activities related to plant growth and development, food and nutrition in space, and the effects of gravitropism and phototropism. It is already being used in dozens of educational centers and is available for sale online.

In addition to the APB and the Space Garden, ORBITEC created another commercial product as a result of its involvement with NASA's bioresearch. This product, the Biomass Production System for Education (BPSe) is a large, ground-based version of the original BPS. Commercialization of the growth system began with a Phase III SBIR contract, "Biomass Production Education System (BPES)," which developed and delivered several classroom plant growth systems to NASA in June 2004. Under a second Phase III contract, the hardware design was updated from the original, delivering 10 more growth chambers, up markedly from the original 4, and 12 more sensor packages. Furthermore, ORBITEC is investigating design and manufacturing options for reducing the unit costs as well as developing learning modules that utilize the BPSe in classroom settings.

ORBITEC is currently proposing that the outcome of its work with NASA on the VEGGIE be sent into orbit, providing continuous crops of fresh produce to the ISS crews. The company is also making plans to distribute the BPSe through programs that form partnerships among public schools, institutions of higher learning, and private enterprises, with the goal of increasing the quantity and quality of engineering, math, and science students. ❖

With its unique bellows system and hydration septum, the Space Garden is already being used in dozens of education centers.

Bringing Thunder and Lightning Indoors

Originating Technology/ NASA Contribution

Piezoelectric materials convert mechanical energy into electrical energy and electrical energy into mechanical energy. They generate electrical charges in response to mechanical stress and generate mechanical displacement and/or force when subjected to an electric current.

Scientists at Langley Research Center have developed a piezoelectric device that is superior in many ways to those that used to be the only ones commercially available. It is tougher, has far greater displacement and greater mechanical load capacity for a comparative voltage operation, can be easily produced at a relatively low cost, and lends itself well to mass production.

The NASA-developed piezoelectric device is also unique in that it is more efficient in extracting electrical energy from the mechanical energy that goes in. It works on a simple principle. A thin ceramic piezoelectric wafer is sandwiched between an aluminum sheet and a steel sheet and held together with LaRC-SI, an amorphous thermoplastic adhesive with special properties created by NASA at Langley. The sandwich is heated in an autoclave, and the adhesive melts. When the sandwich cools, the adhesive bonds the parts together into one piezoelectric element. While they cool, the components of the element contract at different rates, since they are made of different materials. This differential shrinkage causes the element to warp in either a convex or concave shape, depending on which way it is oriented. The shrinking of the outside metal layers places the inside piezoelectric ceramic under mechanical stress. If the element is cantilevered by clamping one side and then plucked, it reverberates like a diving board that has just ejected a diver.

This way, a small amount of mechanical energy can result in a relatively long period of electrical generation. When the piezoelectric element is used for the creation of electricity, it is called Lightning.

This same sandwiched piezoelectric wafer can also convert electrical energy into mechanical energy. Then, it is called Thunder. Electricity goes in, excites the element, and then, mechanical energy in the form of movement is generated.

Partnership

Face International Corporation, of Norfolk, Virginia, holds several licenses to the Langley piezoelectric technology, including the patent on LaRC-SI and the exclusive international marketing rights. Face is now manufacturing a commercial version in mass quantities with its manufacturing partner, Sunnytec Company Ltd., at a new plant in Taiwan.

Face International Corporation has a manufacturing plant in Kaohsiung, Taiwan, where it mass produces the Thunder and Lightning piezoelectric components.

When completely assembled, the Lightning Switch looks much like a typical garage door opener.

Product Outcome

The first mass application of this piezoelectric technology is Face International's Lightning Switch. The Lightning Switch is a wireless, batteryless, remote-controlled light switch, a way to install or replace light switches without any new wiring and without batteries. It is certified for use in the United States and Canada.

Test marketing of the Lightning Switch product started rather humbly last fall, with three mall kiosks in Hampton Roads, Virginia, and an Internet site devoted to the device.

During the test marketing, the product was also aimed at holiday shoppers who might want a remote switch for turning on and off Christmas lights. Holiday revelers who plugged the lights in behind the tree and would otherwise have to move mounds of gifts could now turn the lights on and off without having to brave the tinsel.

The Lightning Switch consists of a remote control transmitter that is modeled after a standard European light switch and a receiver that either plugs into an electrical socket or is wired into an electrical junction box. Pushing the button on the remote control generates enough electricity to send a coded radio signal to the receiver to switch on whatever is plugged or wired into the receiver.

Holiday sales at the kiosks were promising, and Internet sales also contributed to the early success, but these were essentially a marketing experiment for Face International. Serious efforts to penetrate the North American market are underway during this second half of 2005 as Face International begins offering the Lightning Switch for sale through electrical supply houses.

Although it was, in part, marketed as a device for turning on and off Christmas lights, the customers have found many additional, clever uses for the Lightning Switch. The majority of people have used it to install, replace, or rewire lighting controls without the hassle and cost of knocking holes in the walls and ceilings, or having to hire an electrician. The Lightning Switch installs in minutes and can save hundreds of dollars per switch in rewiring costs.

A popular use of the Lightning Switch is in leased or rental properties, where certain tenants may want a switch in one spot, while the next resident may want to have the switch elsewhere. With this device, both can be accommodated, and with no added expenditure by the landlord.

Other uses that customers have found for the device include a taxi-calling system for hotel bellmen; as a call-for-assistance system in assisted living facilities, nursing homes, and hospitals; a control lift for the elderly or disabled; and a signal for a casino table dealer to call for drinks or additional chips.

In addition, it has been used as a notification system for doctors to indicate to nurses when they are ready for the next patient, to trigger lights on the end of a boat dock, as a safety alert for factory floors, in foot switches for wireless tattoo guns, and to control heating, ventilating, and air-conditioning elements.

Customers have found it helpful for controlling landscape lighting, fountains, and pumps for ponds, and as a safe electrical device in wet areas, such as by pools or hot tubs. Some have even planned to employ it for grounds security, with the transmitters packaged to be put in the ground, on doors, gates, and entryways, for permanent wireless and batteryless intruder alerts.

Design-oriented entrepreneurs at retail establishments have used the Lightning Switch as a control for store fixture lighting, while art collectors have used it for backlighting framed pieces. So, while it was being marketed in malls as a Christmas light switch, consumers saw even more potential.

During this test marketing phase, Brad Face, Face International president, had even larger plans. He was in negotiations to have a manufacturing plant erected to meet the growing need for this technology in additional

The Lightning Switch mounts anywhere and requires no wiring.

applications. The new plant, in Kaohsiung, Taiwan, opened in February 2005. It manufactures and assembles the Lightning Switch products as well as Lightning and Thunder piezoelectric elements. The manufacturing lines mainly consist of machinery that was designed and built for the express purpose of making these products. It gives Face International the capacity to produce 30,000 Lightning or Thunder piezoelectric elements, 30,000 Lightning transmitters, and 100,000 receivers each month. The capacity can be increased in increments of 30,000 by adding work shifts or duplicating the manufacturing line. With this capability, the company is prepared to respond to any demand.

There is a large demand developing for these products and not only in North America. Currently, Face International is in contract negotiations with housing development contractors in South Africa, where the Lightning products have the potential to save builders millions of dollars annually. Houses can be assembled quicker without electrical wiring to the switches, and at considerable savings of skilled labor and materials. The Lightning Switch can then be used to install switches in houses after construction.

Beyond the Lightning Switch, Face International has other applications of the NASA-invented piezoelectric element in development. Using the Thunder version of this piezoelectric product, Face International is

Worker in a cleanroom at the plant in Kaohsiung, Taiwan, where the piezoelectric elements are manufactured. The plant has the capability to produce tens of thousands of pieces per month.

working on improving hearing loss assessment technologies. Assessment of hearing loss is normally conducted by testing for minimum sound level detection. There are two forms of tests used for the basic evaluation of auditory function. The first, air-conduction testing, involves presenting precisely calibrated sounds to the ears, usually by routing the signals through headphones to the external ear canal. The second, bone-conduction testing, sends precisely calibrated vibrations through the bones of the skull to the inner ear system. Stimulation is received at the skull by placing a transducer either on the mastoid region behind the ear to be tested or through transducer placement on the forehead.

There has been a long-standing problem inherent in the construction and function of bone-conduction transducers used in auditory testing. Typically, these devices have been restricted in the usable frequency range, particularly above 4000 Hertz, and they have been limited in the amplitude with which sound can be presented to the skull. Bone-conduction transducers have relied on electromechanical components to generate the vibrations. Such transducers do not operate in a linear manner, and, as a result, individual audiometers must be calibrated to the idiosyncratic properties of the bone-conduction transducer to be used with that system. A further problem arises when the transducers are used on a daily basis. When dropped, the transducers frequently break or alter their output characteristics.

Researchers at the Hearing Center of the Hollins Communications Research Institute (HCRI), in Roanoke, Virginia, have been working on development of a new audiometric system for hearing assessment. They have partnered with Face International to create a new bone-conduction transducer that would overcome the major shortcomings of traditional transducers. The new transducers are the correct physical size, with the desired frequency range, linear operation across the relevant range, significant increases in power levels, and they come in a rugged package. The new HCRI/Face International

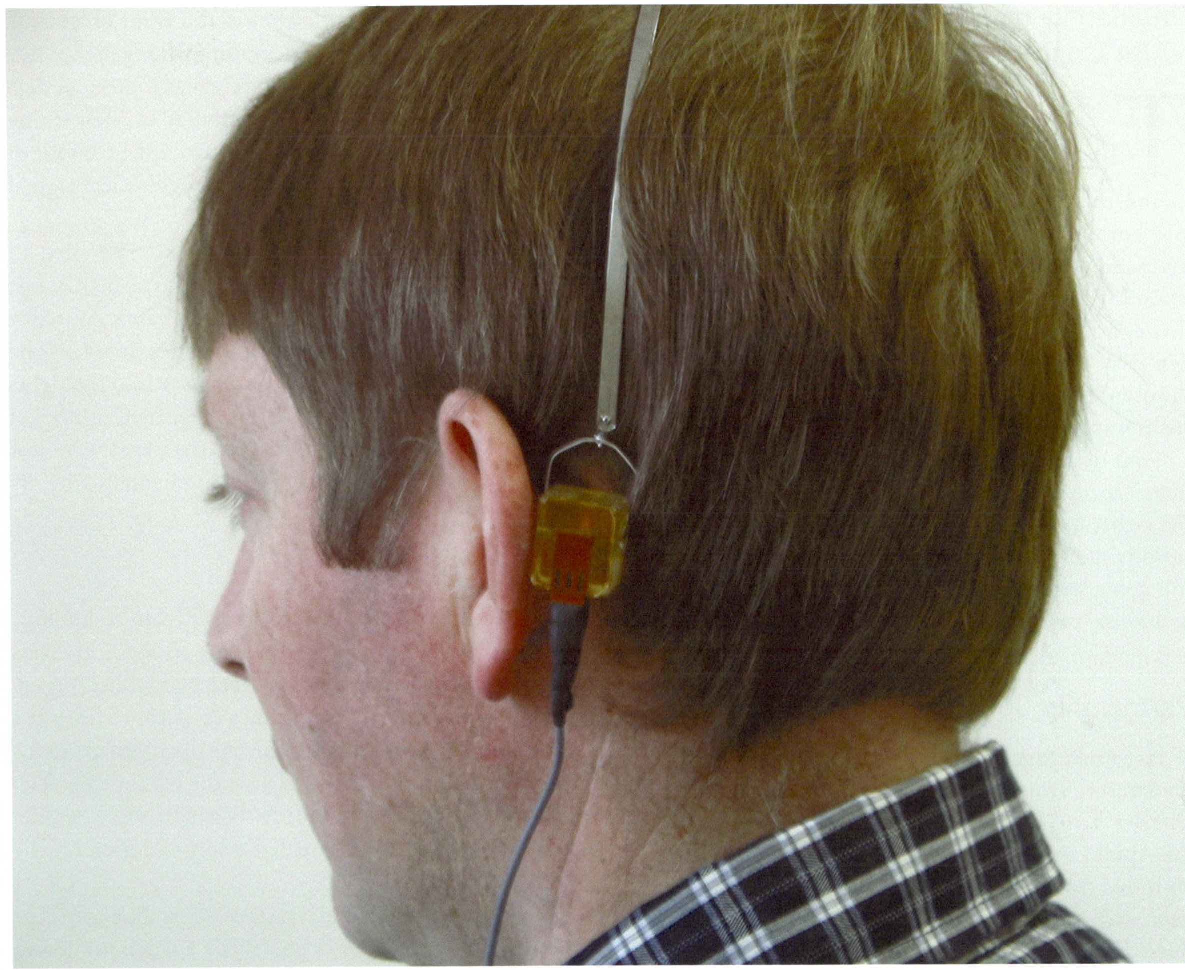

Face International Corporation has partnered with the Hearing Center of the Hollins Communications Research Institute to create durable and accurate hearing test equipment using the NASA piezoelectric technology.

bone-conduction transducers hold up to daily clinical use and even passed the informal stress tests of being dropped on the floor repeatedly.

Brad Face alludes to other applications, most of which are still in their infancy. But with the ability of the new manufacturing plant to create as many of the elements as he could need, and the myriad uses customers are finding for the technology, the applications are limitless. ❖

Lightning® and Thunder® are registered trademarks of Face International Corporation.
Lightning Switch™ is a trademark of Face International Corporation.

Forty-Year-Old Foam Springs Back With New Benefits

Originating Technology NASA Contribution

The most recognized and widely used NASA spinoff is at it again. Temper foam, whose origins date back to 1966 when it was developed to absorb shock and, thus, offer improved protection and comfort in NASA's airplane seats, has paid its dividends to Earth repeatedly, and in many different ways. It has padded the helmets of the Dallas Cowboys throughout the 1970s and 1980s, protected bedridden patients from bedsores, and comforted the feet of thousands wearing stylish shoes that incorporate the cushioning material in their insoles.

Four decades later, the world has come to realize that there are no bounds to temper foam's benefits. Though the rights to the technology have been shared amongst various manufacturers, the original product maker is still going strong, pushing temper foam into new arenas, including automotives, amusement parks, prosthetics, and modern art.

Partnership

As an aeronautical engineer with the Systems Dynamics Group at North American Aviation, Inc., Charles Yost* helped to build a recovery system for the Apollo command module in 1962. His experience with this recovery system would come in handy 4 years later, as Yost was contracted by NASA, through Stencel Aero Engineering Corporation, in Asheville, North Carolina, to assist in the improvement of airline seating for crash and vibration protection, and the development of energy-absorbing techniques for increased survivability. During this time, Yost created an open-cell, polymeric "memory" foam material with unusual viscoelastic properties; that is, it possessed both high-energy absorption and soft characteristics.

NASA's Ames Research Center fit this cushion-like material into a new airplane seat design that not only offered better impact protection in the event of an accident, but enhanced passenger comfort on long flights, due to an even distribution of body weight and pressure over the entire contact area. Initially referred to as "slow spring back foam," the material flowed to match the contour of the body pressing against it and returned to its original shape once the pressure was removed.

The outcomes of Yost's research were detailed in a NASA report entitled "Human Survival in Aircraft Emergencies." In 1969, he formed Dynamic Systems, Inc., to sell the technology as "temper foam." The Leicester, North Carolina-based company sold the rights to the technology in 1974, but later returned to market second- and third-generation derivatives that were less temperature-sensitive and more environmentally friendly than the original version.

Product Outcome

Dynamic Systems expanded temper foam technology from applications for aircraft seating to medical cushions and seating systems for the severely disabled. Eighty percent of the company's product sales come from the medical industry, where its SunMate open-celled, viscoelastic cushioning material is used for orthopedic seating pads, mattress pads, and the Foam-In-Place Seating (FIPS) system, a custom-molded seating solution for the severely disabled.

In addition, Dynamic Systems offers Pudgee. Like SunMate, Pudgee is an open-celled, viscoelastic foam; SunMate can support weight, whereas Pudgee's main function is to compress and conform, therefore providing relief to help alleviate skin breakdown. In other words, Pudgee generates less friction and pressure on tender areas of the skin. Pudgee also transfers moisture, wicking it away from the body.

For the ultimate balance of comfort and support, Dynamic Systems bonded the SunMate and Pudgee materials to create a cushion it calls Laminar. The Pudgee layer on top flows to alleviate pressure and the SunMate layer underneath provides distributed support.

While Pudgee is extremely soft and squishy, SunMate is available in several different formulations, from

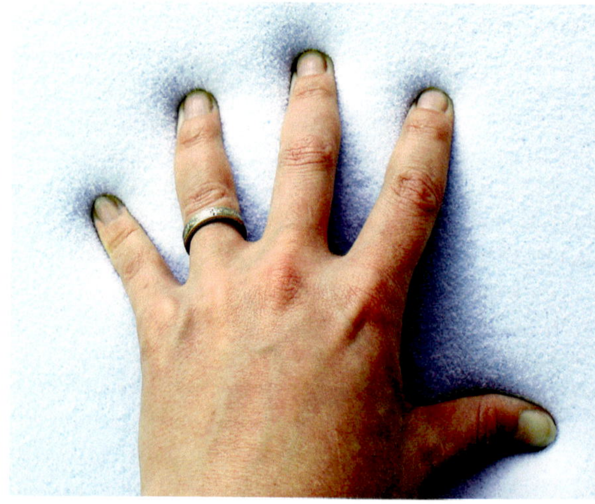

Initially referred to as "slow spring back foam," temper foam matches any pressure against it and slowly returns to its original form once the pressure is removed.

extra-soft to extra-firm. All products form Dynamic Systems are available for customization, and the company offers technical support throughout the application development and implementation stages.

The company's FIPS custom-molded seating system—featured prominently in *Spinoff* 1988—is intended for individuals with complex positioning needs, due to deformities or excessive muscle tone problems. Used primarily for wheelchairs, a FIPS-contoured insert provides advanced posture control by gently holding a person in position. By distributing seating pressure evenly, FIPS improves blood circulation and relieves pressure from pressure points. In short, FIPS helps slow progressive deformities, improves body functions, and eases the soreness and fatigue associated with confined or prolonged seating.

Producing a FIPS insert is an easy process that takes less than a day and does not require expensive equipment. The process involves liquid SunMate "Foam-In-Place" ingredients, which are mixed and poured into a molding bag by medical specialists who ensure that the resulting mold conforms to the patient's body and wheelchair to provide the most therapeutic body position. Within minutes, the liquid ingredients set. All that is left is trimming of any excess material and covering. For patients with less-severe debilitations where complex contouring is not required, Dynamic Systems offers a fabric finish molding bag that is pre-upholstered and ready for immediate use after the FIPS insert sets.

With slight modifications, the FIPS process is well-suited for prosthetic cosmesis. Prosthetic limbs incorporating SunMate materials have the natural feel of flesh and can be molded into a variety of muscle tones and dyed to match exact skin color. Otto Bock Healthcare PLC, a global supplier of innovative products for individuals with limited mobility, utilizes FIPS to develop prosthetic arms. The company's design is specifically intended for amputees missing an entire arm from the shoulder joint.

Once cured, the SunMate foam prevents friction between the shoulder skin and the prosthesis, creating a soft, comfortable interface. Furthermore, the open-celled structure of the foam allows it to breathe, preventing heat buildup.

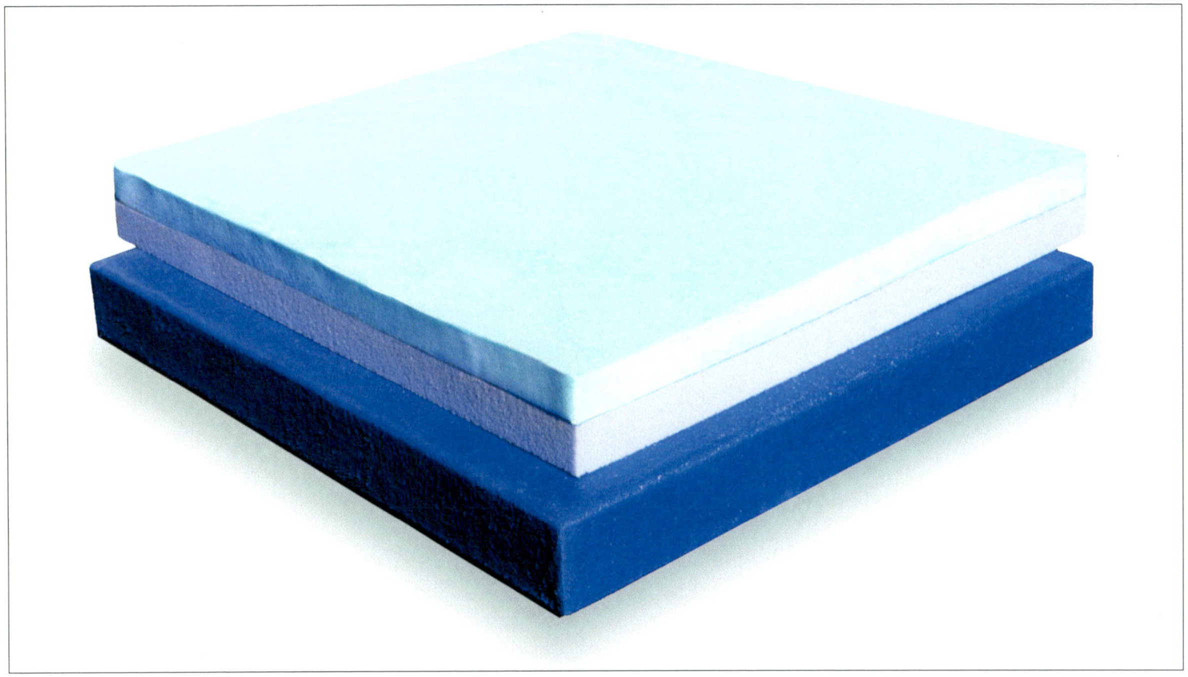

The Laminar cushion, a combination of SunMate and Pudgee materials, provides the extra pressure relief needed to counteract extreme soreness and fatigue. Customers can select from a wide range of cushion thicknesses and pressure supports to create a combination tailored to their requirements.

The same medical advances applied to the treatment of humans can be easily transferred to veterinary applications. When a large animal suffers a leg injury that requires bracing or splinting, healing can be a long, slow process. Using a layer of cushion material beneath an orthotic brace can speed up healing, however. Equine Prosthetics, Inc., of Ocala, Florida, uses SunMate padding in splints, braces, and prostheses for dogs and horses, including its "signature" horse, Thor.

Thor is a 16-year-old, thoroughbred gelding with a right hind foot prosthesis. The majority of his leg is present, although completely nonfunctional from the fetlock down. An accident had caused trauma to the leg, severing his deep and superficial flexor tendons and his lateral suspensory, and breaking his lateral splint bone. Equine Prosthetics rescued him with the intention of putting him to sleep, because his condition was so unusual and extreme, but the company learned that he could be saved and kept comfortable if fitted with a special, customized prosthesis.

Equine Prosthetics needed a molding system that could accommodate complex contours and still provide impact resilience. FIPS met its requirements, mending Thor's leg so that he can now buck, play, and graze, happily and healthily. The horse's condition has improved so much that he now travels to children's hospitals to visit with children who are facing amputation and prosthetic replacement of their own limbs. The goal is to help these children see that it is indeed possible to overcome such an obstacle.

The FIPS technology is also a catalyst to new viscoelastic foam applications extending far beyond the world of medicine.

In stock-car racing, a driver once said he used to feel like he had been "beaten to death" after races. Now that the driver's vehicle is outfitted with a FIPS shock-absorbing insert, he says a race now feels like "a trip around the block."

FIPS technology is now used in NASCAR, Formula 1, Champion Auto Racing Team (CART), and Indy Racing League racecars. One NASCAR driver, whose vehicle featured a custom FIPS insert, walked away from a serious crash unscathed.

Performance Analysis, a motorsports engineering and design firm that manufactures seat safety systems, uses FIPS in professional cars for, among other properties, its low level of toxic emission in the event of a fire. The SunMate product line traditionally does not contain fire retardants, but Dynamic Systems offers a fire-retardant additive for applications in which regulations require the material to pass specific tests. The additive does not detract from the SunMate cushion's impact energy-absorption properties.

Besides FIPS, other SunMate materials are being used in racecars. In Formula 1, SunMate materials have been selected for head and neck supports, due to their comfort, light weight, and versatility.

Motorcycle saddles are also being refurbished with Dynamic Systems's viscoelastic foam technology. The company is offering bikers hours of newfound riding comfort, since motorcycles are typically deficient in shock absorbers and lumbar supports. In 1998, it advertised comfortable, fatigue-reducing materials for "do-it-yourself" motorcycle-seating retrofits, in Rider magazine. In just 3 months, it received 161 responses to its advertisement.

According to Dynamic Systems, motorcycle saddles are easy to refurbish with either SunMate or Laminar cushioning. Motorcyclists who have had their seats refurbished with these foams have reported reduction and elimination of backaches, stiffness, and numbness that result from prolonged riding. Additionally, they maintain that the improved saddles relieve them of pressures induced by seams and rivets on jeans, as well as items in pockets, such as wallets and change.

The SunMate and Pudgee technologies, as well as FIPS, are additionally being used to soften a saddle of a different kind. Salvisa, Kentucky-based Equine Environmental Consulting, Inc., has developed a highly specialized line of saddle pads and saddle-fitting products for racehorses using Dynamic Systems's foam materials. In addition, the Master Saddlers Association, of Woodbine, Maryland, and County Saddlery, Inc., of Lisbon, Maryland, have used SunMate shims under saddles to correct saddle fit and alleviate back and shoulder soreness in horses. The foam padding protects the horse's skin from rubbing against saddle leather, which can cause raw spots or calluses. Once a saddle fits the horse correctly, improvement becomes evident in the animal's skin, joint health, and behavior, leading to improved scores in competition. Yet another organization, Toklat Originals, Inc., of Lake Oswego, Oregon, used FIPS and SunMate for saddle shims that correct asymmetry and other irregularities in a horse's back. SunMate is also used to line the starting gates at racetracks, as well as areas in and around stables, to protect the often easily excitable horses from injury.

Dynamic Systems also dabbles in the arts with its foam products. SunMate and Pudgee cushions have appeared in several art exhibits featuring new and unusual materials applied in unconventional ways, from major shows at The Museum of Modern Art and the Smithsonian Cooper-Hewitt National Design Museum, to student exhibits at the Harvard Graduate School of Design, the Rhode Island School of Design, and the Royal College of Art.

At the Smithsonian Cooper-Hewitt National Design Museum, Pudgee was chosen for display in the "Skin: Surface, Substance, and Design" exhibit, due to its skin-like properties, the benefits it provides for skin, and the sensation it evokes upon feeling it for the first time. At the Rhode Island School of Design, a student created a contemporary lounge chair that incorporated a ball of Pudgee foam, for a course whose objective was to create a piece of furniture featuring novel or unconventional materials. The ball doubled as a spring-like structural support and a seat cushion. Other pieces of contemporary furniture have been

Dynamic Systems, Inc.'s foam technology buffers bumps and bruises, making it a good fit for motorcycle saddles.

designed by art students, including a Pudgee-stacked bench and a Pudgee sofa lounge. Dynamic Systems's staff of project coordinators assists artists and designers with the technical details of transforming their cushion concepts into works of art.

In fashion and design, a gown-tailoring service and dress design company by the name of Design Studios is using full-size body casts made from FIPS foam to create customized dresses. The owner of Snohomish, Washington-based Design Studios started the company to fulfill consumer demand for custom dress forms that accommodate different sizes and shapes, and reduce the number of time-consuming fittings and alterations required for making tailored clothing. The owner experimented with a variety of pourable molds to create body casts, but quickly discovered that once set, they would either be too soft, too rigid, too powdery to work with, or otherwise incapable of withstanding the constant pinning and manipulation that are essential in tailoring.

Soon thereafter, the owner learned about the FIPS process from a nearby engineering laboratory. She contacted Dynamic Systems, and technicians helped her determine the appropriate formulations to successfully create body molds to fit her needs. According to Dynamic Systems, the molds turned out to be so accurate, that clients need only show up for two fitting sessions.

In recreation, U.S. Aqua Sport, Inc., of Arvada, Colorado, has designed an inflatable "bumper" raft that will not sink if punctured. The circular-shaped raft, intended for whitewater rapid attractions at theme parks, features an internal chamber filled with SunMate foam. This layer of cushioning enhances the safety of amusement park patrons who choose to confront the rugged rapids.

Recently, Dynamic Systems has seen increased interest in use of the firmest formulation of SunMate for archery targets. The material absorbs severe impact, weathers well, and is durable enough to withstand thousands of shots, allowing for longer use of a target. Another advantage of this material is that it is practically self-healing. Arrows

SunMate, Pudgee, and Laminar benefit both rider and horse, whether it be with a more comfortable riding saddle or a form-fitting prosthetic leg for injured animals.

from an 85-pound pull bow may penetrate up to 8 inches and be removed by hand with little or no effort (arrows are very difficult to remove from most target materials) and with very little damage to the material.

The firmest formulation also provides significant bullet impact pressure reduction when used as a backing in bulletproof vests and as a shock absorber to protect personnel in military vehicles that run over landmines.

Coming full circle to the initial NASA application, SunMate and Pudgee materials are making their way into the seats of private and commercial aircraft, as well as into the seats of helicopters. In Hendersonville, North Carolina, two helicopter pilots had a harrowing experience when their helicopter crash-landed because of engine failure. Fortunately, both pilots walked away with only minor injuries, due in part to the special high-impact absorption of the Dynamic Systems cushions in their seats.

The impact-absorption capabilities of SunMate make the material a good match for ejection seating in military aircraft. Recent tests by the U.S. Air Force showed that a 2-inch SunMate cushion performed "substantially better" than other proposed ejection seat cushions in reducing the force of impact in the catapult stage.

Spinning back to NASA, the Space Agency recently built a demanding obstacle course, placed over a special 4-inch-thick SunMate cushion floor, to evaluate astronaut post-flight performance.

Astronauts who return from space experience difficulty with balance and walking as they re-adapt to gravity on Earth. The special floor provides an unstable support surface that increases the challenge for astronauts regaining their equilibrium. Their physiological responses are monitored as they try to negotiate their way through the tricky obstacle course.

A total of three SunMate floors have been built for this purpose, one of which was sent to Russia for use by the astronauts returning from the International Space Station.

The list of applications and products goes on and on, as does Dynamic Systems's commitment to producing a low-cost, highly advanced cushion material for comfort and energy absorption. In the last 10 years, the company has been honored as an inductee in the U.S. Space Foundation's Space Technology Hall of Fame, presented with a certificate of recognition from NASA, and has received the prestigious "Governor's Cup" award as the top small business with less than 50 employees in the Southeast, all for inventing temper foam.

With all of these honors and a growing customer base, Dynamic Systems is sitting pretty comfortably these days. ❖

**In memory of Charles Yost, who passed away on March 29, 2005.*

SunMate® is a registered trademark of Dynamic Systems, Inc.

Experiments With Small Animals Rarely Go This Well...

Originating Technology/ NASA Contribution

In the mid-1950s, a young U.S. Air Force engineer named Clark Beck began work with what is now one of NASA's most prolific spinoffs, the radiant barrier technology. Beck's work involved creating materials that could withstand the immense heat created by passage through the Earth's atmosphere. He was working on structures and resources that could withstand the fluctuations in temperature created by a skip reentry, where a craft would "skip" along the surface of the atmosphere, gradually making inroads sufficient for reentry, a process that took the craft from extreme heat to frigid cold every few seconds. The material also needed to withstand millions of pounds of pressure per inch of bending without twisting, the simulated force of reentry. Without reflective material, the craft would get what Beck refers to as "red hot wings," and without the required flexibility, the craft would break apart.

One result of Beck's work was the discovery of the useful properties of radiant barrier material. The Space Agency used Beck's design work for the materials that went into building the space capsules, heat resistant instrument panels, and, in conjunction with the Air Force, an early spacecraft prototype, the DynaSoar, that looks remarkably similar to the present-day Space Shuttle.

NASA used the thin, shiny, silver material to protect the first space explorers from the harsh environment of space, which could range from -460 °F to 541 °F. If the engineers had used conventional insulation for the space suits, the fabric would have been 7-feet thick, a little awkward for even the most nimble of astronauts to maneuver. Radiant barrier technology was clearly the solution. It reflected the astronaut's body heat back into the suit to keep him warm, while at the same time reflecting radiant energy from the Sun outward to keep him cool.

The radiant barrier material reflected more than 95 percent of the radiant energy away from the wearer, while tiny holes in the fabric allowed moisture to escape and longer heat waves to get through. This amazing fabric had an added benefit that made it ideal for its space-bound application: It weighed only 17 pounds per 1,000 square feet.

NASA has used this material ever since the Gemini and Apollo missions, on virtually all of its spacecraft, and even on unmanned missions as thermal protection for instruments.

The Hutch Snuggle keeps the pet's hutch cool in the summer and warm in the winter.

Partnership

NASA's radiant barrier technology has been public domain for the past 3 decades. It is used in "space blankets" and countless other applications, including energy-saving home insulation, awnings and canopies, agricultural insulation, automotive insulation, and protective apparel, the likes of which is used by firemen. These are only a few of the many applications found for this material.

Scratch and Newton, Ltd., a company with worldwide Internet sales, employs this technology to improve the lives of pet rabbits and guinea pigs. Experiments involving small, cute, furry animals almost always end badly for the critters. In this instance, though, science and cute animals can coexist.

Product Outcome

Scratch and Newton has a clever premise: Scratch, a fumbling rabbit, has a slew of problems, and Newton, a genius guinea pig, tinkers at his workbench to solve them. The "solutions" are the company's products. They include hutch covers, run covers, and bottle wrappings. The ideas and products actually come from company founders Chrissie Slater and Lydia Ross, two Britons who had been experimenting with bubble wrap, thick unplasticised polyvinyl chloride, and other materials that were hard to work with and had significant problems, like breathability and condensation. After extensive experimenting, they decided that, as Slater says, "The radiant barrier is just ace!"

The Hutch Snuggle protects the outdoor pet rabbit or guinea pig from the elements. Made of the radiant barrier material, it helps reduce summer heat gain and winter heat loss, making the animal's environment more comfortable. The rabbit or guinea pig gets a comfortable shelter from the harsh elements in the winter and a cool patio in the summer. The Hutch Snuggle even features a clear window for the pet to look out and the keeper to look in. Used in conjunction with the Hutch Hugger, a hutch cover made of weather-resistant fabric, the pet is guaranteed to be safe from wind and rain. All of these items come in different sizes, so they should fit just about any hutch. Custom-made hutch covers, however, are also available.

These little pets cannot spend all of their time locked in hutches, though. They need to stretch and exercise. For this purpose, Scratch and Newton also makes the Run Hugger, a radiant barrier cover for the animal's play pen. This way, even when the weather gets especially rough, the animal can still have some much-needed daily play time.

Frankie and Munroe are fur-ever pleased to have their fresh water protected by the Bottle Snug.

The Bottle Snug, another product developed by Scratch and Newton for the small critters, is designed to keep a pet's water supply fresh and drinkable. The radiant barrier bottle cover prevents water from freezing during winter months and keeps water cool and fresh during the hotter seasons. This also cuts down on the green algae that can sometimes grow in water bottles in warmer weather.

Scratch and Newton has sold its products around the world, though it plans to target Canada, home to many chilly bunnies, and New Zealand, land of the overheated pet guinea pigs. It plans to proceed with its line of thermal critter-care products and look to NASA technology to find other innovations that may be of use.

Clark Beck, the engineer who pioneered the radiant barrier technology, is now retired, though still active in his local Dayton, Ohio, engineers club, and not surprised by the widespread use of the radiant barrier technology. He has seen it used in myriad ways. "Its uses are only limited to one's imagination," he says. "It can be used wherever you find heat or cold." ❖

Hutch Snuggle™, Hugger™, Run Hugger™, and Bottle Snug™ are trademarks of Scratch and Newton, Ltd.

NASA, the Fisherman's Friend

Originating Technology/ NASA Contribution

Every angler has his secrets, whether it be an old family recipe for stink bait, a midnight worm-hunting ritual, or the most coveted of all, the no-fail fishing hole. Most of these secrets are lore and legend, passed through generations, and coveted more than the family's best tableware. Each of these kernels of wisdom promises the fisherman a bite at the end of the line, but very few are rooted in fact and science.

There is one, though....

NASA partnered with a company on the bayous of Mississippi and Louisiana to use satellite data to create a marine information system, a space-age fish finder.

This product provides up-to-date information about the location of a variety of fish, including yellowfin tuna, bluefish, blue marlin, white marlin, sailfish, blackfin tuna, little tunny, and swordfish. The system shows peaked catch rates, and may be the only true fish-finding product on the market.

Partnership

Digital Media, Inc. (DMI), teamed with fellow Gulf Coast native, NASA's Stennis Space Center, through Mississippi's Space Commerce Initiative and tapped into a wealth of satellite imagery and weather-forecasting tools essential for guiding a successful fishing trip.

DMI is a television production company specializing in videos for outdoor enthusiasts, including broadcasts on hunting, shooting, golfing, camping, local attractions, and of course, fishing. Recently, they have focused energy and attention on geospatial products for the broadcast industry.

Product Outcome

The DMI fish-finding service combines remote sensing data with proprietary software and can provide

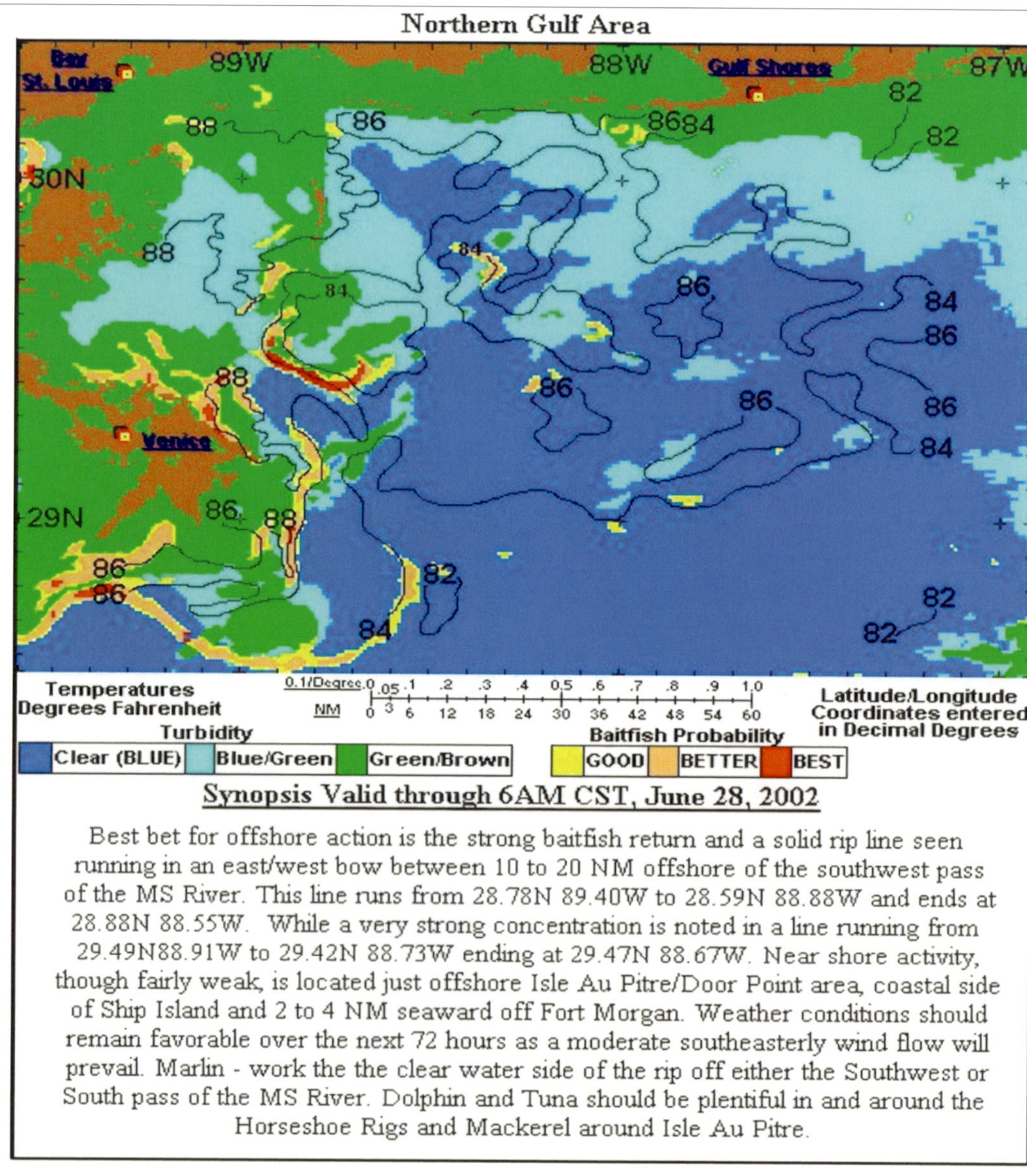

This easy-to-read display can be broadcast directly to an onboard computer, telling the boat's captain exactly where to go to catch the most fish.

information about the entire coastal United States, the Gulf of Mexico, the Caribbean, and Central America. In the past, marine informational services have been isolated to specific geographic regions.

DMI collects satellite information about water surface temperature, temperature gradients, water clarity, and depth, then uses a mathematical algorithm originally developed and tested by the National Marine Fisheries Service that accurately identifies conditions where bait fish are likely to congregate. Using Global Positioning System coordinates, DMI can direct fishing parties straight to these spots. According to DMI owner Vincent Phillips, this "is the only service to include the three primary pieces of the puzzle that are necessary to accurately pinpoint the concentrations of sportfish and their associated forage."

The fish-finding service merges the water temperature, turbidity data, and the proprietary algorithm into an easy-to-read printout. Fishermen can purchase this readout as a daily forecast, a seasonal product, or an annual subscription. According to Phillips, "Marine information of this type traditionally has been very expensive and consequently used by the few who could afford it." This service, however, is a fraction of the price of other fish-finding services, and the only one to offer the complete picture.

Although the fish-finding service takes the guesswork out of the trip, the angler will still have other challenges to reckon with.

The service provides 24-hour access to images and text of marine information. The images are available via Internet in full color or in black and white for facsimile transmission, and a fisherman or boat captain can have a fishing map sent directly to the deck of his boat. These images are not only for the experienced fisherman, though; the readouts are clear enough that even the novice fisherman should be able to decipher them.

Fishermen who use the fish-finding service save time and money. Since they know where the fish are, they need only motor out to the exact location. It would normally take hours of guesswork to tool around on the water trying to find a productive location for fishing. Fishermen drop the line, wait, drop the line, wait, and then move to the next spot. This wastes valuable fish catching time, and with the prices of commercial fishing ventures and gasoline considered, this advent saves everyone money, as it reduces the time that the boat is being operated unproductively. The saved fuel has obvious environmental benefits.

This product appeals to a wide variety of marine interests, and it has found users among sportsmen, commercial fishermen, sailors, divers, charter boat captains, and private boaters. DMI also provides printouts to special interest groups, including saltwater tournament competitors and marine researchers conducting experiments with tideland, near-shore, and deep water conditions.

Broadcast and print media can also subscribe to these services and then provide general maps to accompany their weather reports. This product shows viewers general trends in their locations, and is a great tool to encourage involvement by locals and visitors alike in coastal activities, events, and sports. ❖

Crystal-Clear Communication a Sweet-Sounding Success

Originating Technology/ NASA Contribution

"That's one small step for man, one giant leap for mankind."

On July 20, 1969, millions were glued to their television sets when NASA astronaut Neil Armstrong offered these famous words via live broadcast, upon becoming the first man to ever step foot on the Moon.

This historic transmission was delivered from Armstrong's headset to the headsets of Mission Control personnel at NASA, and then on to the world. To ensure that this message was delivered loud and clear to all, NASA collaborated with private industry to deliver the best headset-communication technology possible. Today, the heart of this technology beats in a high-end line of lightweight and wireless solutions for homes, offices, contact centers, and dispatch centers.

Partnership

During the late 1950s and early 1960s, most commercial and military pilots were employing communications headsets that proved to be clunky and cumbersome. A number of these pilots changed to hand-held microphones in place of the uncomfortable headsets; but as high-performance jet planes evolved, the headset became essential as a safety feature.

In a move to make pilot headsets more uniform and comfortable, United Airlines put out an open solicitation in 1961, inviting anyone who was interested to tender their ideas for an improved design. Courtney Graham, a United Airline pilot, was one of the many who thought the heavy headsets should be replaced by something lighter. Graham was interested in submitting a new design concept to United Airlines, so he recruited a friend named Keith Larkin to be his business partner. Larkin had been working for a small company called Plane-Aids, importing spectacles and sunglasses that contained tiny transistor radios from Japan.

Their original design was inspired by these radio-equipped glasses. They submitted this concept to United Airlines for review, though they quickly learned that it would not be feasible for use, as not all pilots wore glasses or preferred wearing sunglasses while flying. Consequently, Graham and Larkin came to find that a headset using a lightweight headband would be more practical. For the communication aspect of the headband, Larkin added an acoustic tube to two very small transducers commonly used in hearing aids, for use as both a microphone and a receiver.

United Airlines accepted this headband design, which immediately evolved into the MS-50 headset. Graham and Larkin's successful solution motivated them to start a company of their own, Pacific Plantronics, Inc. (now called Plantronics, Inc.).

In 1962, the newly formed, Santa Cruz, California-based company introduced the MS-50 to the aviation industry as the world's first lightweight headset. United Airlines started ordering the new headsets for its pilots shortly thereafter, and the Federal Aviation Administration adopted a modified version for air traffic control; United Airlines is still using the MS-50 today.

Graham and Larkin soon found out that it was not just the aviation industry that was experiencing problems with the uncomfortable headsets of the time—astronauts, telephone operators, and call centers, too, had similar issues.

NASA was looking to outfit its astronauts with a lightweight, self-contained radio after it had lost contact during a dire situation. Lieutenant Colonel Virgil Ivan "Gus" Grissom was the command pilot of 1961's Liberty Bell 7 (Mercury-Redstone 4) mission, America's second manned space flight. Following the space mission, the Liberty Bell 7 capsule had a successful reentry into Earth's atmosphere and experienced a relatively mild impact when it landed in the Atlantic Ocean.

The capsule righted itself after splashdown and all seemed well, but minutes later, while Grissom was preparing for rescue by a helicopter scheduled to retrieve him and the capsule, its explosive hatch blew prematurely, causing water to gush in and flood the cabin.

According to the pre-established plan, the explosive hatch was not to be blown until the capsule had been lifted clear from the water. Because the hatch blew too early, Grissom was forced to exit the flooding Liberty Bell 7, thereby abandoning all of the spacecraft's radio communications equipment, and since the helmets of astronauts were not equipped with radio communications in the early days of space flight, he was unable to make contact with his recovery team.

Grissom was able to free himself and get the visual attention of the helicopter circling above for safe rescue, though the recovery team had no choice but to part with

A U.S. Marine helicopter lifts astronaut Virgil Ivan "Gus" Grissom to safety, following splashdown of the Liberty Bell 7. The spacecraft had just sunk below the water.

the spacecraft, due to the added weight of the water that had flooded it.

NASA did not want to risk losing radio contact with an astronaut again, so it contracted with ITT Labs, in Fort Wayne, Indiana, to develop a self-contained radio system. ITT Labs delivered to NASA a radio transceiver called a Kellorad unit, which happened to feature Pacific Plantronics's MS-50 headset.

When NASA astronaut Walter "Wally" Schirra first saw the lightweight headset component of the Kellorad unit, his interest—and curiosity—peaked. He contacted Pacific Plantronics and asked the company whether the MS-50 receiver could be implanted directly in his astronaut helmet. Pacific Plantronics was confident that it could satisfy Schirra's request, so NASA awarded the company a contract to do so.

Pacific Plantronics assembled its Space Environmental Communications (SPENCOM) division to begin working on a reliable solution. SPENCOM personnel traveled to NASA's Manned Spacecraft Center (now Johnson Space Center) and Kennedy Space Center to meet with and get design feedback from Schirra and several other astronauts, including Gordon Cooper. (Schirra and Cooper were two of the seven original NASA astronauts.) Based on the feedback provided, Plantronics and NASA decided to focus on a noise-canceling microphone that used dual, miniature transducers. A helmet receiver using the miniature transducers was also developed. The SPENCOM group worked in conjunction with Manned Spacecraft Center engineers to design and implement the new noise-canceling microphone and receiver.

SPENCOM and NASA had a working microphone design with the correct impedance needed for space communications within 11 days, and Schirra was the first to use the new communication technology during the Sigma 7 (Mercury-Atlas 8) mission.

Unfortunately, ground control lost contact with Schirra for approximately 7 critical seconds during the "Max Q" phase of liftoff, when the most noise was introduced into the capsule, because aerodynamic pressure was at its highest. Changes had to be made to compensate for this noise level, so the design team added a dual-tube, noise-canceling device to the microphone. The transmission using the new noise-canceling technique provided greatly improved communication.

The improved unit was developed for the Faith 7 (Mercury-Atlas 8) mission, the follow-up mission to Sigma 7, commanded by Cooper, and continued to be successfully used through the Gemini, Apollo, and Skylab missions.

Pacific Plantronics gained increasing recognition as a result of participating with the Space Program, as the advancements made for astronaut communication flourished in other markets. By year 6 of operation, the company's sales reached $5.4 million, with 100,000 headsets in use. The Plantronics of today credits its collaboration with NASA and NASA's use of the jointly developed microphone technology for this instant market credibility.

Plantronics, Inc.'s CA10CD cordless push-to-talk headset and amplifier gives call center agents and other business professionals room to roam.

Product Outcome

Plantronics's entire line of communications headsets and telephony solutions leverages the company's expertise in mission-critical headset design and development for NASA. Its CA10CD cordless push-to-talk headset and amplifier serve as the perfect solution for contact centers, extending cordless mobility to the entire line of Plantronics professional headsets. Recognized as a Certified Space Technology by the U.S. Space Foundation, the CA10CD offers hands-free operation for up to 150 feet away from the base unit, dual volume control for optimum sound and clarity, remote-call answering, and a rechargeable battery that allows 8 hours of continuous talk time. Whether it is picking up the latest fax across the hall or checking reference materials on the other side of the building for a customer, the CA10CD gives call center agents and other business professionals room to roam.

The CA10CD also made its way into radio dispatch centers at police, fire, taxi cab, waste management, and utility company facilities. Additionally, NASA uses this product in a variety of applications, including refueling, field testing, and vehicle-assembly and range-management operations.

In the office setting, the Plantronics CS50 Wireless Headset System enables a hands-free workstyle that suits the needs of today's busy professionals, enhancing their freedom and efficiency. The sleek, self-contained wireless CS50 addresses the number one product concern of business professionals who have chosen not to use a tethered headset: they do not want to be tied to their desks or have a wire in the way (based on a Plantronics study). With the CS50, office professionals can communicate up to 300 feet away from their desk phone, for up to 8 hours. The lightweight headset also uses built-in digital encryption so conversations are secure.

The CS50's wireless mobility also gives businesses an important competitive advantage. Most workspace is unoccupied for 60 percent of the workday, and every day more than 70 percent of business calls are missed (UNWIRED, 2003). This means that valuable time is lost exchanging voicemails. The CS50 effectively reduces phone tag, letting users not only roam hundreds of feet from their desks, but, with the addition of the optional HL10 handset lifter, answer and end calls remotely. Since the CS50 is as intuitive and easy to use as answering the phone, the learning curve is minimal.

"In a fast-paced, people-powered business like ours, the phone is a powerful and indispensable tool. The CS50 has greatly improved my department's accessibility and productivity," said Adam Keen, vice president of special projects for MGM and United Artists. "The CS50's combination of mobility and comfort enables us to multitask much more effectively. Also, unanswered calls are now a rarity, which means there's less chance we'll miss out on something important."

CS50 users can satisfy their individual style and comfort preferences and gain additional convenience and functionality with optional accessories. Beyond the HL10 handset lifter that provides one-touch call answer/end, there is a behind-the-head neckband, which provides another style option for personal comfort, and a wall- or desk-mountable in-use indicator that lets others know a user is on a call.

The GameCom Halo 2 Edition Xbox headset combines an innovative design with advanced microphone technology, resulting in a stable, great-sounding headset. The in-the-ear wearing style provides optimal fit and boom placement, which lets users hear and be heard clearly, even in the loudest game environments.

On the home-consumer front, Plantronics technology is enhancing the overall experience for video game enthusiasts. In August 2004, the company announced that it had shipped more than 1 million headsets to the Microsoft Corporation for the Xbox Live online service, which lets Xbox video game console owners around the world communicate with and play against each other, 24 hours a day. Equipped with noise-canceling microphones and a swiveling boom, the Plantronics headset delivers superior audio so that gamers can take advantage of multiplayer, voice-enabled games by strategizing with teammates or taunting opponents. The headset also features a compact Flex-Grip in-the-ear design for comfort during extended play. Plantronics offers similar headsets for personal computer gaming.

"We see tremendous opportunity in the gaming market and believe that the popularity of the Xbox Live headset is just one indicator of how successful we can be in growing our leadership position," said Matt Miller, director of Plantronics product marketing. "With the combination of increased adoption of online gaming through services like Xbox Live and our headset's crystal-clear audio, we're able to deliver the most intense gaming experience possible."

At the 2005 International Consumer Electronics Show, the largest consumer technology show in the world, Plantronics was the only headset maker to win a "Best of Innovations" award. The company was recognized for innovation in design and technological advances for its new Voyager L510S Bluetooth headset system. (Bluetooth is a standard developed by electronics manufacturers that allows any sort of electronic equipment—from computers and cell phones to keyboards and headphones—to make its own connections, without wires, cables, or any direct action from a user.)

The Voyager L510S multipoint system can simplify communication for a wireless headset, creating seamless switches between landline office phones and voice-enabled Bluetooth mobile phones, laptops, and personal digital assistants. At approximately half an ounce in weight, it began shipping in spring 2005.

In July, Plantronics introduced the Pulsar 590, a Bluetooth stereo headset that lets users listen wirelessly to music and movies with excellent audio quality and switch seamlessly to mobile phone calls with the single touch of a button. The Pulsar 590 expands Plantronics's family of award-winning Bluetooth products to provide consumers with a broad range of headsets to meet individual lifestyles. ❖

Xbox® is a registered trademark of Microsoft Corporation.
Xbox Live™ and GameCom™ are trademarks of Microsoft Corporation.
Flex-Grip® is a registered trademark of Plantronics, Inc.
Voyager™ and Pulsar™ are trademarks of Plantronics, Inc.
Bluetooth® is a registered trademark of Bluetooth SIG, Inc.

The lightweight and versatile Plantronics Voyager Bluetooth headset lets business professionals go wherever their conversations take them. It delivers all-day wearing comfort, clear voice transmission, and superior sound quality, all in an easy-folding design that slips into a pocket or purse for quick storage. With up to 6 hours of talk time, users can roam up to 33 feet from their Bluetooth devices.

Inertial Motion-Tracking Technology for Virtual 3-D

Originating Technology/NASA Contribution

In the 1990s, NASA pioneered virtual reality research. The concept was present long before, but, prior to this, the technology did not exist to make a viable virtual reality system. Scientists had theories and ideas—they knew that the concept had potential—but the computers of the 1970s and 1980s were not fast enough, sensors were heavy and cumbersome, and people had difficulty blending fluidly with the machines. Scientists at Ames Research Center built upon the research of previous decades and put the necessary technology behind them, making the theories of virtual reality a reality.

Virtual reality systems depend on complex motion-tracking sensors to convey information between the user and the computer to give the user the feeling that he is operating in the real world. These motion-tracking sensors measure and report an object's position and orientation as it changes. A simple example of motion tracking would be the cursor on a computer screen moving in correspondence to the shifting of the mouse. Tracking in 3-D, necessary to create virtual reality, however, is much more complex. To be successful, the perspective of the virtual image seen on the computer must be an accurate representation of what is seen in the real world. As the user's head or camera moves, turns, or tilts, the computer-generated environment must change accordingly with no noticeable lag, jitter, or distortion. Historically, the lack of smooth and rapid tracking of the user's motion has thwarted the widespread use of immersive 3-D computer graphics.

NASA uses virtual reality technology for a variety of purposes, mostly training of astronauts. The actual missions are costly and dangerous, so any opportunity the crews have to practice their maneuvering in accurate situations before the mission is valuable and instructive. For that purpose, NASA has funded a great deal of virtual reality research, and benefited from the results.

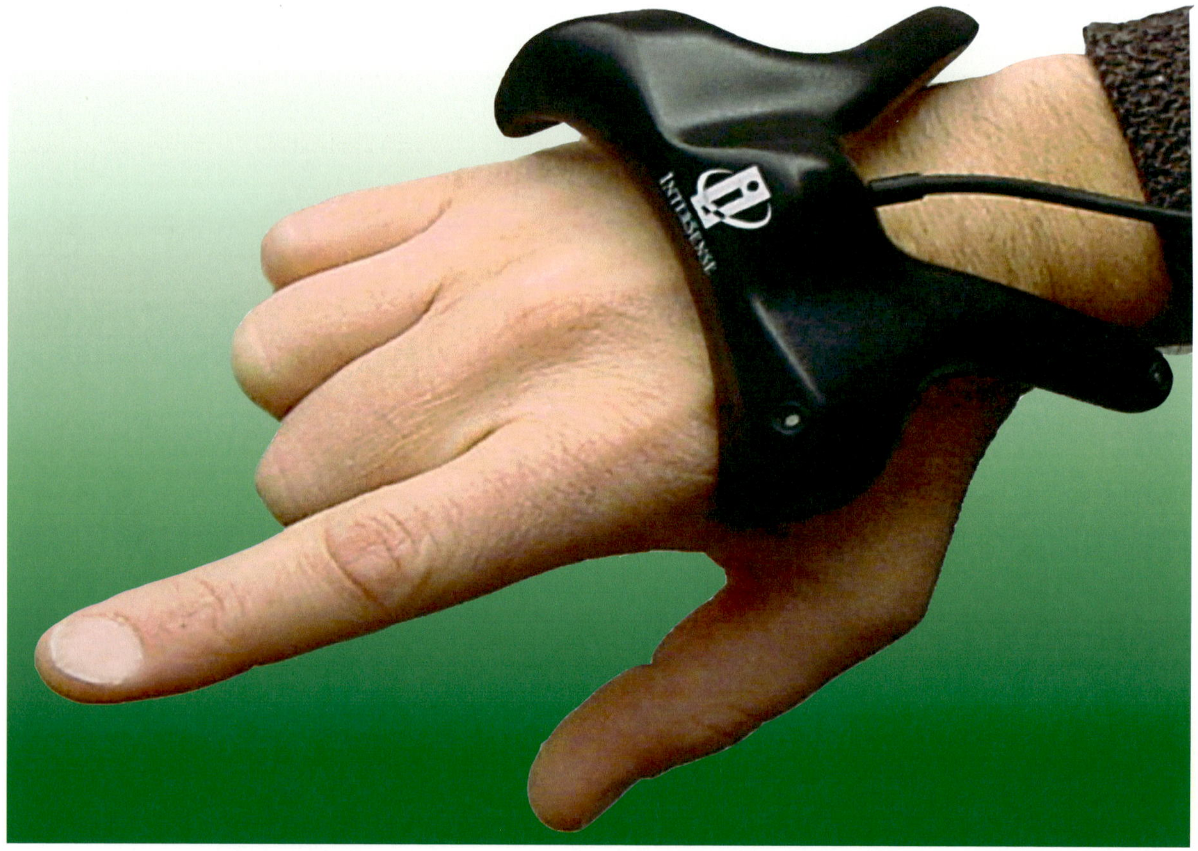

Virtual reality systems depend on motion-tracking sensors to relay information between the user and the computer. Pictured here is the InterSense IS-900 MiniTrax 6-DOF Hand Tracker for immersive environment interaction.

Partnership

Scientists at Ames, led by Elizabeth Wenzel, were looking for a better way to track head motion for use with 3-D audio systems. Dr. Wenzel's group was conducting research experiments on binaural localization with subjects wearing tracked 3-D audio headsets. Traditional headsets were either too slow, had high latency, or were too bulky. They were looking for a better headset.

A graduate student at the Massachusetts Institute of Technology, Eric Foxlin was conducting research in virtual reality systems when he came up with the idea to develop a miniature tracking device based on the same technology found in large, ship-borne navigation tracking devices. The Inertial Measurement Unit (IMU) uses large precision gyroscopes, accelerometers, and compasses to estimate, with great accuracy, the position and orientation of ships and aircraft. Foxlin used this concept to build a miniature IMU using low power, low cost microelectromechanical systems (MEMS) components with support from an Ames seed grant. The first miniature, body wearable, sourceless tracker was developed and used as a technology seed to attract venture capital and start InterSense, Inc., in 1996.

Bedford, Massachusetts-based InterSense now develops precision, miniaturized inertial motion-tracking technology extensively deployed in simulation and training, entertainment applications, clinical and medical settings, for oil and gas discovery, virtual design and testing, and in a variety of research applications. A cutting-edge virtual reality company, one of its frequent customers is NASA.

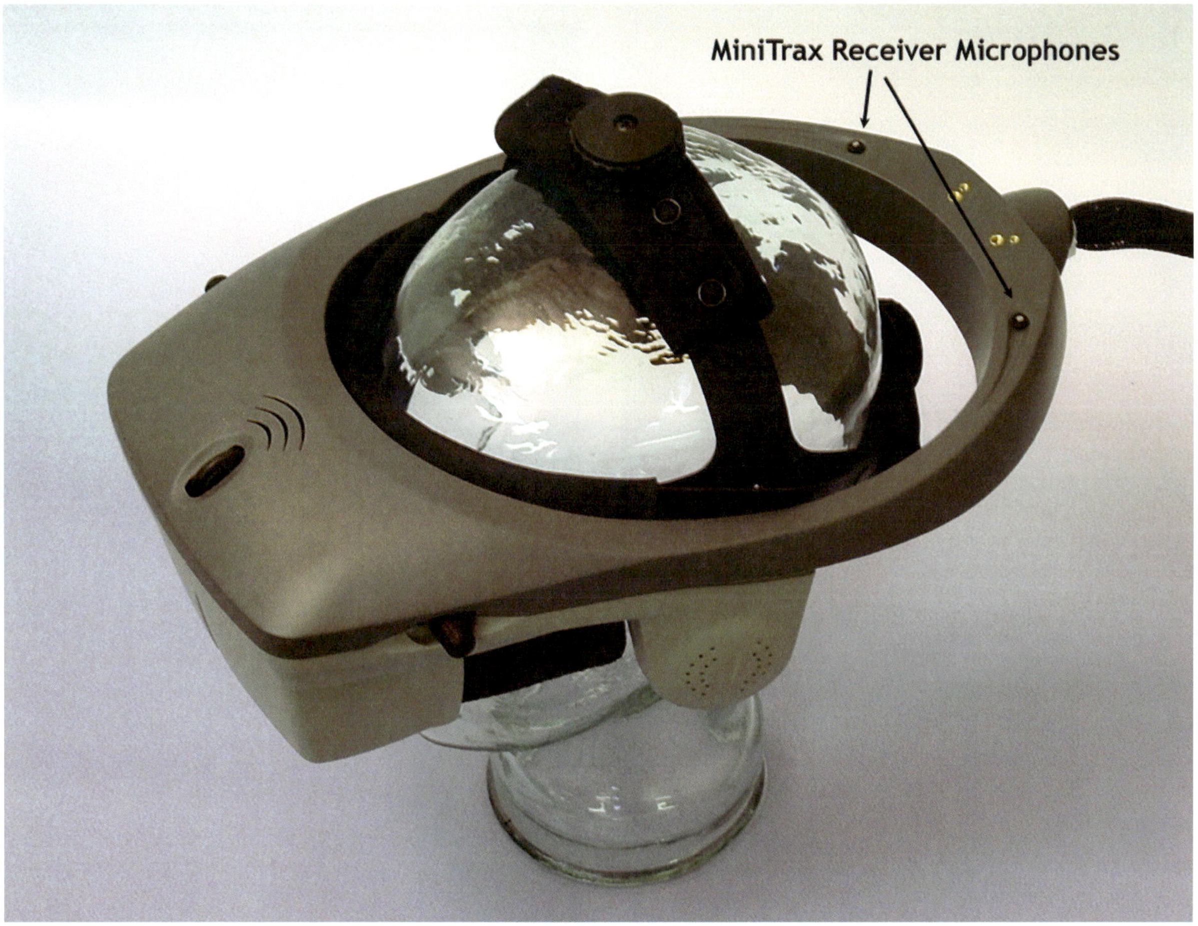

InterSense's immersive headsets, like this NVIS Head Mounted Display, bring 3-D to life when integrated with IS-900 MiniTrax 6-DOF Inertial-Acoustic Tracking Technology.

InterSense's IS-900 Virtual Environment Tracking system was recently installed at Glenn Research Center for use in a 3-D virtual immersive display wall. Combined with 3-D fluid flow analysis software, the stereo display wall is used to gain better understanding of 3-D flow modeling of jets, turbines, and Space Shuttle aerodynamics. Simulated failure analysis is also performed, such as what happens when a jet turbine blade is impacted by a foreign object.

Product Outcome

InterSense's unique technology tracks motion with an unmatched combination of size, cost, and precision. Its users create products and applications that allow anyone—from product designers, computer and Internet game developers and players, to scientists, teachers and students, assembly line workers, and video and film production companies—to interact with virtual 3-D images just as they do physical objects.

The InterSense products have the added bonus of allowing the users unlimited range, negligible jitter, high update rates, and low latency. The units have no discernable interference or line-of-sight problems and offer superior motion prediction.

InterSense offers standard products which provide orientation-tracking and location-tracking systems. Each product offering has a unique combination of inertial technology which is sensor-fused with complementary tracking technologies. The InertiaCube product family combines MEMS gyros and accelerometers with magnetometers to offer an accurate, sourceless tracking sensor. The IS-900 product family fuses the MEMS inertial components with an ultrasonic positioning system to offer an accurate tracking solution for wide-area tracking with wireless tracking devices. The IS-1200 product family fuses MEMS inertial components with passive or active optical position references, providing an autonomous tracking solution for mobile or moving vehicle applications.

Virtual Design Review in University of Central London's Immersive Display Room, using the InterSense IS-900 Wand and Head Trackers.

With its extensive product line, InterSense is the leading provider of head and helmet tracking systems to major government contactors like the Boeing Company, Lockheed Martin Corporation, and L3 Communications, Inc.'s Link Simulation and Training division (Link).

One of the projects it worked on with Link is the Aviation Combined Arms Tactical Trainer (AVCATT) now employed by the U.S. Army for training helicopter pilots. It allows multiple users, so pilots can practice simulated flights as a fleet, as opposed to individually. The AVCATT is the Army's newest aviation training simulator which includes a series of networked simulators offering a dynamic, reconfigurable system used for collective training and mission rehearsal. The AVCATT provides up to five functional cockpits, and even allows pilots to choose between different models currently employed by both the Army and the National Guard.

Link and InterSense have also teamed together to create the F/A-18C Distributed Mission Training System, a project similar to the helicopter simulator. This, however, is used to model jet aircraft, and, like the AVCATT, it allows pilots to train simultaneously in real-time, 3-D missions in simulated cockpits.

Another project using InterSense's NASA-funded technology is the Stinger Missile Trainer, a 40-foot dome with projected terrain and aircraft images. The dome allows the trainees a full 360-degree surround scene and 70-degree vertical field of view. The system trains up to three

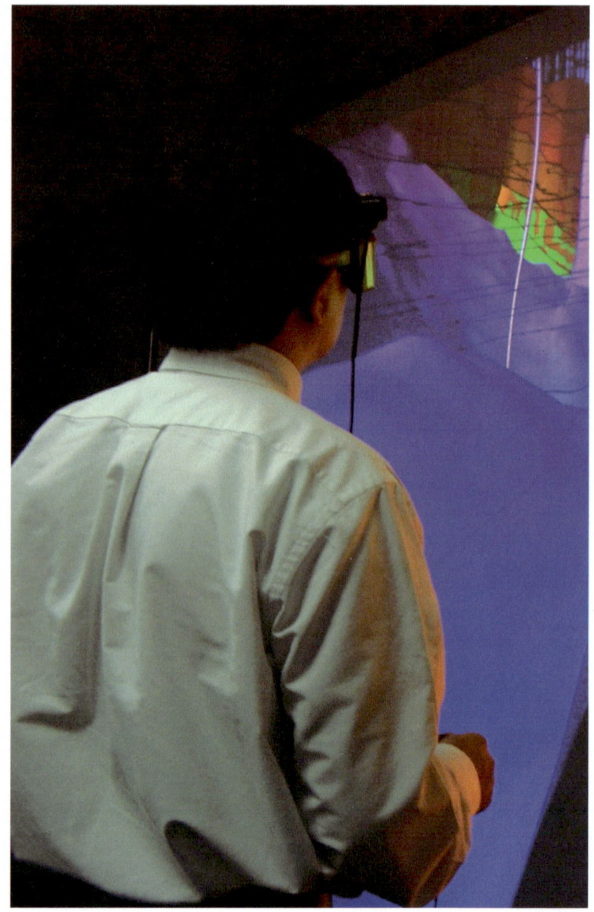

The immersive virtual reality systems that InterSense works on can be found virtually everywhere.

gunners, individually or simultaneously, to identify, acquire, and track airborne targets, and then launch a Stinger missile. An upgraded version of this program was recently developed for domed simulators to improve the overall training efficiency and realism. As part of the upgrade, InterSense worked with several other cutting-edge simulation companies to free the trainees from tethers and cables that were connected to training weapons.

Another InterSense partner, Fakespace Systems, Inc., of Marshalltown, Iowa, constructed a reconfigurable visualization system as part of an immersive environment simulator used to study how soldiers use equipment in combat zones at the Army Research Laboratory (ARL). The RAVE II visualization system consists of three detachable, large-scale stereoscopic display units that ARL uses to project realistic imagery of simulated hostile environments. The immersive environment simulator integrates the RAVE II with InterSense's motion-tracking device and an omni-directional display system consisting of three self-contained, rear-projected modules that can be arranged to form a flat wall display or an immersive theater environment. The immersive environment simulator also integrates motion-tracking with an omni-directional treadmill allowing soldiers to run and move in any direction within the virtual hostile terrain in order to simulate live combat conditions.

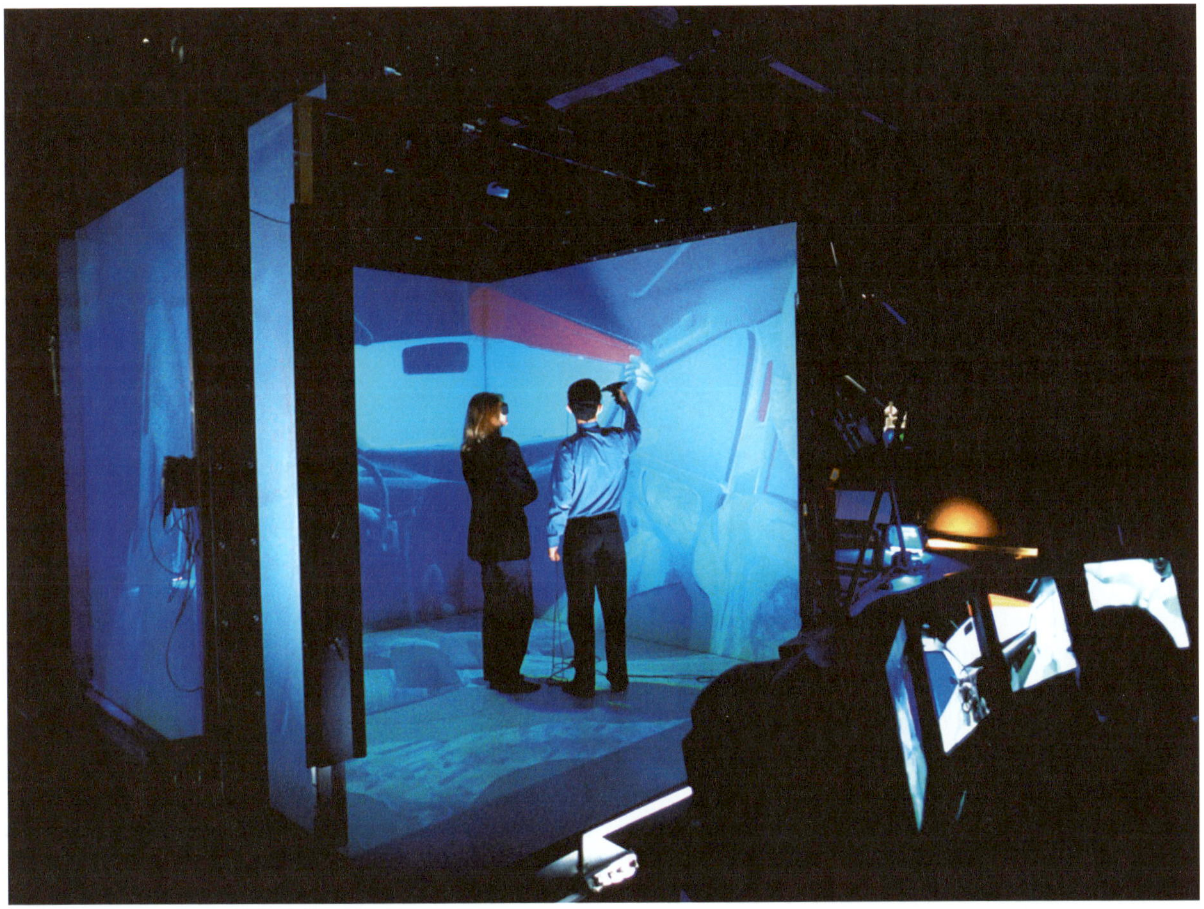

Virtual Design Review of Automotive Interior in Peugeot's Immersive Display Room, where designers use the InterSense IS-900 Wand and Head Trackers.

It is not just the military that is benefiting from InterSense's work. The company worked to integrate its inertial head tracker into the Microsoft games, Flight Simulator 2002 and Flight Simulator 2004: A Century of Flight. Touted by Microsoft as "the world's most popular game," Flight Simulator sold over 21 million copies. In the game's "Virtual Cockpit" mode, the user becomes fully immersed in the cockpit and views the world as a real pilot does with a full 360-degree view. Along with the consumer market, numerous military departments and pilot training schools utilize this program as a cost-efficient training simulator for beginning pilots.

InterSense's inertial motion-tracking technology can also be found in hospitals, where it is used to supplement ultrasound imagery; in oil fields, where it helps workers locate new locations for wells; at General Motors vehicle design facilities; television and film studios; and in university research. Its motion-sensing work, funded by NASA, can be found virtually everywhere. ❖

Then Why Do They Call Earth the Blue Planet?

The DOLPHIN "swims" behind research vessels and samples the water to validate satellite imagery. It provides valuable information about ocean color and quality.

Originating Technology/ NASA Contribution

While the most common photographs of Earth taken from space show the planet covered in blue water, NASA has managed to produce detailed color images, using satellite imagery, that show the remarkable variation of colors that actually make up the oceanic surface. An ocean's color is determined by the interaction of surface waters with sunlight, and surface waters can contain any number of different particles and dissolved substances, which could then change the color.

The particles are mostly phytoplankton—the microscopic, single-celled ocean plants that are the primary food source for much marine life. Remote detection of phytoplankton provides information about the uptake and cycling of carbon by the ocean through photosynthesis, as well as the overall health of the water. Inorganic particles and substances dissolved in the water also affect its color, particularly in coastal regions. Satellite images can be used to calculate the concentrations of these materials in surface waters, as well as the levels of biological activity. The satellites allow a global view that is not available from ship or shore.

NASA's orbiting satellites offer a unique vantage point for studying the oceans. By resolving the biological, chemical, and physical conditions in surface waters, they have allowed the oceanographic community to make huge leaps in its understanding of oceanographic processes on regional and global fronts. The study of ocean color, in particular, has been integral in helping researchers understand the natural and human-induced changes in the global environment and establishing the role of the oceans in the biochemical cycles of elements that influence the climate and the distribution of life on Earth.

These views provide researchers with a great deal of data that have a vast range of applications. The first level of data describes the optical properties of the water, such as the transmission and scattering characteristics. From this data, a second level of derived products is generated that describes what is in the water, including concentrations of chlorophyll (the primary photosynthetic pigment), particles, and dissolved substances. These derived products are used in research on global warming, ecosystem composition and health, fisheries research, pollution monitoring, coastal eutrophication (very high levels of phytoplankton growth, which lead to dangerously low oxygen levels), seasonal variability, sediment transport, river outflows, currents, and the regional influences of weather patterns and monsoons. Satellite data are invaluable for studying global phenomena and phenomena occurring over large regions, like El Niño.

The first observations of ocean color from space were carried out by the Coastal Zone Color Scanner (CZCS), which operated on NASA's Nimbus-7 research satellite from 1978 to 1986. This groundbreaking research provided scientists with an unprecedented view of the world's oceans and opened wide this field of oceanographic study. The CZCS was the longest running observational satellite of this type, and spawned a series of other satellite missions with the specific goals of adding to this wealth of knowledge.

Partnership

Through a Phase II **Small Business Innovation Research (SBIR)** contract with the Earth Science Applications Directorate at NASA's Stennis Space Center, WET Labs, Inc., of Philomath, Oregon, developed a new technology for validating ocean color images from satellites. It calls the device the Diving Optical Profiler and High-speed Integration Network, or DOLPHIN.

Product Outcome

DOLPHIN is a much-needed research system that can provide high-quality field data to validate data from ocean color satellite imagery.

Validation of satellite images is done by collecting 1) in-water optical data to validate the level one data described above and 2) discrete water samples to validate the level two data with conventional laboratory analyses. The traditional method for sample collection involved lowering instruments from boats at stations set apart at periodic intervals. The problem with this method of sampling for validation work is that the ocean is dynamic in space and time. A more synoptic method of sampling is required that is more consistent with the synoptic views obtained by remote aircraft and satellites. In the time expended setting up for sampling and relocation for conventional vertical profiling, water shifts and the aircraft or satellite data become dated.

With DOLPHIN, a single research vessel tows the sampling unit while it undulates vertically in the surface waters. Optical parameters are measured on the underwater vehicle while water for conventional laboratory analyses of biological, chemical, and geological parameters is simultaneously pumped to the surface through a hose in the tow cable. The results are more synoptic in space and time than with the traditional method, since the craft need not stop to gather samples. DOLPHIN provides researchers with a continuous transect of points for validation, as opposed to single stationary points. Gathering all this data for validation of both the level one and level two oceanic properties on-the-fly makes the validation exercise more accurate and efficient. This saves time and money.

The unit can gather data as it bobs and swims like a dolphin behind the boat. The bobbing has features other than making its name appropriate, though, as the raising and lowering of the unit allows sampling of the water at different depths, giving researchers more than just surface conditions. This sampling below the very surface is important, because the ocean can have considerable vertical layering of its biological and chemical constituents, especially in coastal waters. The reflected light seen by the aircraft or satellite comes from a surprisingly large range of depths in the upper ocean, extending as deep as 300 feet, so vertical sampling in surface waters is required for validation.

Besides the validation applications for the DOLPHIN, the package also has the potential to be used for water quality monitoring, to track freshwater in-flows, to map dispersion points of nutrients, hydrocarbons, and other pollutants, as well as track harmful algal blooms. ❖

Casey Moore (WET Labs president), Dr. Richard Miller (NASA Stennis Space Center), and Dr. Michael Twardowski (WET Labs project director) hoist the portable DOLPHIN unit into the research vessel.

Valiant 'Zero-Valent' Effort Restores Contaminated Grounds

Originating Technology/ NASA Contribution

Dense non-aqueous phase liquids (DNAPLs) are chemical compounds that can contaminate soil and groundwater to the point of irreparability. These substances are only slightly soluble in water, and are much denser than water. Because of their solubility, DNAPLs form separate liquid phases in groundwater, and because of their density, DNAPLs sink in aquifers instead of floating at the water table, making it extremely difficult to detect their presence. If left untreated in the ground, they can taint fresh water sources.

Common DNAPLs include chlorinated hydrocarbon compounds such as carbon tetrachloride, chloroform, tetrachloroethylene, and trichloroethylene. Trichloroethylene was used during the early days of the Space Program, as a solvent for flushing rocket engines, and for metal cleaning and degreasing of equipment, electronics, and heavy machinery. As a result, areas of Cape Canaveral's Launch Complex 34—the site of several historic Saturn rocket launches occurring from 1959 to 1968—were polluted with chlorinated DNAPLs.

Through the direction and guidance of Dr. Jacqueline Quinn, an environmental engineer in the Spaceport Engineering and Technology Directorate at NASA's Kennedy Space Center, a biodegradable environmental cleanup technology was developed to reductively dechlorinate DNAPL sources in polluted water at Launch Complex 34. It was important for Kennedy to nip this problem in the bud, in light of the fact that the Space Center is also a National Wildlife Refuge, home to thousands of shorebirds, endangered sea turtles and eagles, manatees, alligators, and diverse habitats that include brackish marshes and salt water estuaries.

The success in remediating this historic launch site has led to numerous commercial applications that are restoring the health of our environmental surroundings.

Partnership

Quinn and three University of Central Florida professors developed the Emulsified Zero-Valent Iron (EZVI) technology, with collaborative support from the U.S. Department of Energy, the U.S. Department of Defense, the U.S. Environmental Protection Agency (EPA), GeoSyntec, Inc., and NASA's **Small Business Technology Transfer (STTR)** Program at Kennedy. EZVI uses iron particles in an environmentally friendly oil and water base to neutralize toxic chemicals.

When speaking of the remediation work, Quinn noted, "What makes this new technology so attractive is that it is relatively affordable and easy to implement. Spinoff [companies] could help clean up polluted areas across the Nation and the world."

The Merritt Island National Wildlife Refuge, which shares a boundary with Kennedy Space Center, is home to diverse habitats and endangered animals, including the bald eagles pictured above. This eagle nest is 1 of 12 active nests on the premises. It is important for Kennedy to dechlorinate dense non-aqueous phase liquid sources in polluted water to protect these animals and area habitats.

To set such a cleanup effort into motion, Kennedy's Technology Transfer Office issued non-exclusive licenses for EZVI to several companies, including Weston Solutions, Inc., and Toxicological and Environmental Associates, Inc.

Product Outcome

Headquartered in West Chester, Pennsylvania, Weston Solutions is a leading environmental and redevelopment firm, dedicated to developing and implementing solutions that safely and cost-effectively address complex remediation challenges. Interestingly, the company was the lead environmental contractor for the Space Shuttle Columbia recovery effort, the largest emergency response ever conducted by the EPA.

Following consummation of the Kennedy licensing agreement, Weston Solutions personnel were trained by NASA, with regard to the techniques for preparing EZVI and applying the patented emulsion technology in the field.

The company and one of its industrial clients recently evaluated the EZVI technology for application at one of that client's sites in Tennessee that was impacted by high concentrations of chlorinated solvents that could not be remediated using conventional technology. Upon scrupulous assessment of site conditions and endorsement from associated regulatory agencies, EZVI was applied to treat a zone of soil located approximately 30 feet below ground surface.

The EZVI emulsion was injected under pressure into the target zone. Field work was conducted with strict attention to operating pressures, the amount applied, and the zone of injection. Further, an application array was established onsite to position the injection and assure even coverage over the contaminated "footprint."

Immediately after treatment and at subsequent intervals thereafter, groundwater monitoring and soil-sample analysis were conducted to track the concentration trends and treatment effectiveness. Concentrations of the

Weston Solutions, Inc.'s Emulsified Zero-Valent Iron remediation technology in process at a client site.

targeted contaminant dropped dramatically following the EZVI injection. Concurrently, concentrations of chloride increased, showing the dechlorination of the contaminant. According to Mike Corbin, technical director of Weston Solutions, "the regulators and the industrial client are very excited about the results so far."

Weston Solutions is currently evaluating other sites—nationwide and in Canada—which contain DNAPLs or near-DNAPL concentrations. The company applied EZVI at a site in North Carolina over the past summer. Through the company's applications of EZVI, additional knowledge is being gained on the performance of the technology in different types of soil and various other hydrogeologic environments. This experience will provide valuable feedback to the NASA inventors. The company has maintained open communication with the NASA team as commercial treatments progress.

Another licensee, Toxicological and Environmental Associates, also received NASA training and insight for creating a stable and reactive EZVI emulsion. The Baton Rouge, Louisiana-based company is facilitating the commercial application of the NASA technology through a spinoff company it formed, called Applied Science & Advanced Technologies, Inc. This will be done in place of traditional pump-and-treat methods, which can require decades of treatment time and operational costs, and steam injection and radio-frequency heating, which are very expensive and can trigger contaminant mobilization. According to the spinoff company, EZVI overcomes these limitations by providing a method that is quick, effective, and cost-competitive.

Applied Science & Advanced Technologies invested approximately 1 year in the commercialization of the manufacturing process to create the emulsion. It is currently capable of producing 2,500 gallons per day and can efficiently expand this ability as product demand increases. Additionally, the company has secured an exclusive agreement with an iron powder manufacturer. This will enable it to control costs associated with the production of the EZVI technology, which, in turn, should bolster its marketability in the remediation industry.

Throughout 2005, Applied Science & Advanced Technologies is scheduled to implement EZVI treatments at contaminated sites in Arkansas, California, Florida, North Carolina, and Texas.

DNAPL contamination sites may include, but are not limited to, those created by dye and paint manufacturers, dry cleaners, chemical manufacturers, metal cleaning and degreasing facilities, leather-tanning facilities, pharmaceutical manufacturers, adhesive and aerosol manufacturers, and government facilities.

It should be noted that DNAPLs are present at 60 to 70 percent of all sites on the EPA's Superfund National Priorities List, a registry of national priorities among the known or threatened releases of hazardous substances, pollutants, or contaminants throughout the United States and its territories.

The EZVI technology represents a major step forward in preserving the vitality of our home planet and its inhabitants. ❖

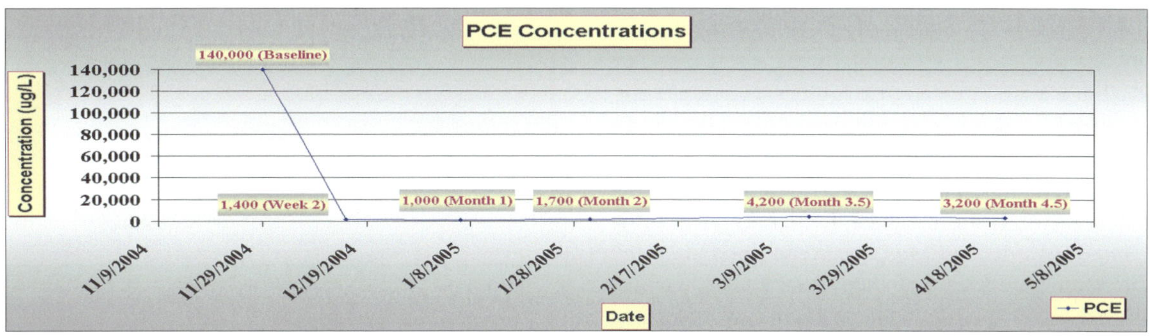

This graph depicts dense non-aqueous phase liquid concentrations before and after treatment was applied to a contaminated site by Weston Solutions, Inc. Concentrations of the contaminant dropped dramatically following injection of the treatment, and, though not depicted here, chloride levels increased, showing that the dechlorination was successful.

Harnessing the Power of the Sun

Originating Technology/ NASA Contribution

The Environmental Research Aircraft and Sensor Technology (ERAST) Alliance was created in 1994 and operated for 9 years as a NASA-sponsored coalition of 28 members from small companies, government, universities, and nonprofit organizations. ERAST's goal was to foster development of remotely piloted aircraft technology for scientific, humanitarian, and commercial purposes. Some of the aircraft in the ERAST Alliance were intended to fly unmanned at high altitudes for days at a time, and flying for such durations required alternative sources of power that did not add weight. The most successful solution for this type of sustained flight is the lightest—solar energy. Photovoltaic cells convert sunlight directly into electricity. They are made of semi-conducting materials similar to those used in computer chips. When sunlight is absorbed, electrons are knocked loose from their atoms, allowing electricity to flow.

Under the ERAST Alliance, two solar-powered technology demonstration aircraft, *Pathfinder* and *Helios*, were developed.

Pathfinder is a lightweight, remotely piloted flying wing aircraft that demonstrated the technology of applying solar cells for long-duration, high-altitude flight. Solar arrays covering most of the upper wing surface provide power for the aircraft's electric motors, avionics, communications, and other electronic systems. *Pathfinder* also has a backup battery system that can provide power for between 2 and 5 hours to allow limited-duration flight after dark. It was designed, built, and operated by AeroVironment, Inc., of Monrovia, California.

On September 11, 1995, *Pathfinder* reached an altitude of 50,500 feet, setting a new altitude record for solar-powered aircraft. The National Aeronautic Association presented the NASA-industry team with an award for 1 of the "10 Most Memorable Record Flights" of 1995.

Pathfinder was succeeded by *Helios*, a remotely piloted flying wing aircraft also developed as part of ERAST. It, too, was originally designed and manufactured by AeroVironment with high-efficiency solar cell arrays across the wing, as well as navigation and emergency lights, improved avionics, and more efficient propellers.

The two primary goals of the *Helios* prototype development were to demonstrate sustained flight at an altitude near 100,000 feet, and to fly nonstop for at least 24 hours, including at least 14 hours above 50,000 feet. In 2001, the *Helios* prototype achieved the first of the two goals by sustaining flight above 96,000 feet for more than 40 minutes on a test flight near Hawaii, during which it reached an unofficial world-record altitude of 96,863 feet. On a later flight in 2003, intended to evaluate a hydrogen fuel cell system designed to provide power at night for extended flight duration, *Helios* became uncontrollable, broke up, and crashed into the Pacific Ocean. The mishap resulted from the inability to predict the aircraft's increased sensitivity to atmospheric disturbances, such as turbulence, following all of the vehicle configuration changes.

Partnership

SunPower Corporation, of Sunnyvale, California, created high-efficiency silicon cells for the NASA/AeroVironment *Helios* and *Pathfinder* solar-powered aircraft under the now-concluded ERAST Alliance. These photovoltaic cells were the most advanced silicon-based cells available for terrestrial or airborne applications and, over time, evolved into the mass-produced A-300. The A-300 is an efficient, high-performance, low-cost, single crystal silicon solar cell, with a unique, all back-contact design. It offers up to 50 percent more power than conventional solar cells.

SunPower first became involved with NASA when the Lawrence Livermore Laboratory, in California, transferred a solar aircraft project under the Ballistic Missile Defense Organization program to Dryden Flight Research

The experimental site at NASA's Dryden Research Center consists of two fixed-angle solar arrays and one single-axis Sun-tracking array. Together they produce up to 5 kilowatts of direct current power on a sunny day, which is equivalent to powering two or three average California homes. The Sun-tracking array tilts to follow the Sun using an advanced real-time tracking device rather than normal pre-programmed mechanisms.

Center. SunPower had been making high-efficiency solar concentrator cells for the laboratory. These cells quickly became the solar cells of choice for the prototype aircraft, now named *Pathfinder* and *Helios*. SunPower's cells were selected for the program because of their high-efficiency performance and lightweight design.

Out of these experiences, SunPower gained the confidence to begin making cells on a much larger scale for terrestrial applications. At the behest of NASA, however, SunPower began to drive costs down, while retaining the high-performance efficiency of its products. Eventually, the company was able to bring costs down to be competitive with conventional cells.

At this stage, SunPower partnered with Cypress Semiconductor Corporation, of San Jose, California, which brought the capital resources and expertise in high-volume manufacturing to the business picture. The first ground installation that involved Cypress funding was at Dryden, where SunPower was able to conduct field tests and performance comparisons on its products through a small solar power demonstration site that provided power to one of Dryden's buildings, and several dimensions of insight came from the experience. The first was the progressive evolution of cost reduction. The installation also helped SunPower to identify uses and performance advantages that would enable the company to get more firmly rooted in large-scale commercial applications.

The Dryden solar power demonstration installation includes three systems: a traditional array and two arrays with SunPower cells. One of the SunPower arrays is mounted on a structure tilted to a fixed angle, while another is installed in a "tracker" that follows the courseof the Sun during the day. Staff can review output data and dynamically compare the performance of the side-by-side systems. Together, the systems produce up to 5 kilowatts of direct current power on a sunny day, which is equivalent to powering several average California homes.

According to SunPower, typical commercial-grade solar cells are in the range of 12- to 15-percent efficient at converting sunlight to electricity, while its A-300 cells are at least 20-percent efficient in a ground application, or up to 50-percent better than the older technology. The efficiency improvement is largely due to the routing of cell electrical connections behind the cells, which was required in the original design for the solar-powered aircraft to maximize the limited space available atop the wings.

The demonstration system permits easy comparison of the advantage of tracking over the fixed-tilt array, verifying the effectiveness of models for predicting array output, and the effect of dust on the arrays. Array efficiency is monitored remotely on a computer. Currently, it is providing supplemental power to Dryden's 7,870-square-foot Public Affairs and Commercialization building.

Long-term plans for this project could include construction of a solar "farm" at Dryden that would power up to one-third of the Center's electrical power needs.

Product Outcome

Consumers can also purchase these same solar cells from SunPower, as many people have elected to do. Homes across the country are now being outfitted with the modern, space-age solar power. The solar panels were even featured on the ABC television show, "Extreme Makeover: Home Edition," which highlighted a team of carpenters and designers who built a home for a family in Flagstaff, Arizona. The donated SunPower panels supply a significant portion of that home's electricity needs.

Solar panels appeal to consumers, because they preserve the environment, save money on electricity bills, and insulate them from escalating energy costs. Some customers consider solar energy patriotic power: a local power source that does not rely on imported fuels.

SunPower's solar panels have no moving parts, creating pollution-free electricity with no noise and virtually no maintenance. Furthermore, both SunPower's all-back contact solar cells and its high-tech, high-performance solar panels perform better than most other solar panels—even in overcast or very hot conditions.

In addition to the test facility at Dryden and home-consumer uses, SunPower has recently partnered with SOLON AG, Germany's largest solar photovoltaic module manufacturer, to open the first phase of the world's largest solar electric plant in Bavaria near Arnstein, Germany.

Solarpark Gut Erlasse is a 12-megawatt solar electric plant located in a working agricultural field and can serve the electricity needs of thousands of customers each year. SunPower's solar cells are used in SOLON's "Mover" electricity generation system, designed for deployment in multi-megawatt solar power plants. Movers automatically tilt and rotate during the day to directly face the Sun at all times, similar to the concept utilized at Dryden.

By combining SunPower's industry-leading, high-performance A-300 solar cells with the Mover's ability to maximize daily energy production, each square foot of solar cells can generate up to double the annual energy of a fixed solar array using conventional solar cells. At Solarpark Gut Erlasse, SunPower's cells are used only in about one-third of the Movers, but generate a higher proportion of the energy there, due to their high performance.

Now that SunPower has achieved commercial success with its mass-produced solar cells and modules, the company is working with the U.S. Department of Energy's National Renewable Energy Laboratory on further improvements in module design to reduce costs for solar power. In addition, SunPower is working with several partner companies on specialized, mass-produced, building-integrated solar products that would leverage its A-300 cell's attractive all-black appearance. Roof tiles that include solar cells and can be installed like regular roofing are one example. SunPower is even looking at other consumer products as a possible next step to making solar energy successful in a variety of markets. ❖

Water and Air Measures That Make 'PureSense'

Originating Technology/ NASA Contribution

Each day, we read about mounting global concerns regarding the ability to sustain supplies of clean water and to reduce air contamination. With water and air serving as life's most vital elements, it is important to know when these environmental necessities may be contaminated, in order to eliminate exposure immediately.

The ability to respond requires an understanding of the conditions impacting safety and quality, from source to tap for water, and from outdoor to indoor environments for air. Unfortunately, the "time-to-know" is not immediate with many current technologies, which is a major problem, given the greater likelihood of risky situations in today's world.

Accelerating alert and response times requires new tools, methods, and technologies. New solutions are needed to engage in more rapid detection, analysis, and response. This is the focus of a company called PureSense Environmental, Inc., which evolved out of a unique relationship with NASA.

The need for real-time management and operations over the quality of water and air, and the urgency to provide new solutions, were reinforced by the events of September 11, 2001. This, and subsequent events, exposed many of the vulnerabilities facing the multiple agencies tasked with working in tandem to protect communities from harmful disaster. Much has been done since September 11 to accelerate responses to environmental contamination. Partnerships were forged across the public and private sectors to explore, test, and use new tools. Methods and technologies were adopted to move more astutely from proof-of-concept to working solutions.

NASA was, and continues to be, involved in a number of these efforts, drawing on support from its experience and expertise with complex systems, advanced technologies, and environmental sciences. The founders

PureSense Environmental, Inc., provides customers in commercial and municipal markets with real-time solutions to detect, know, and respond to critical changes in the quality of water and air.

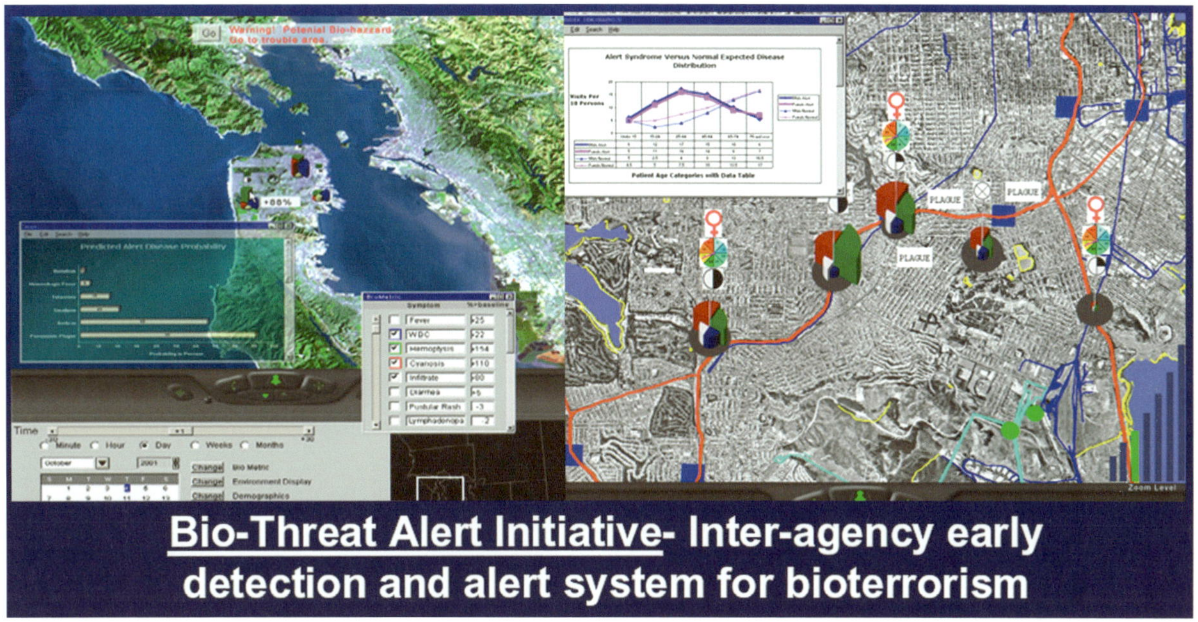

Understanding how to deal more effectively with bioterrorist attacks: PureSense Environmental, Inc., signed two Space Act Agreements in 2003, to support the NASA mission to "protect our home planet," and a Cooperative Agreement in 2004, to collaborate on U.S. Department of Homeland Security projects designed to ensure the security of the Nation's water systems.

of PureSense originally approached NASA to position the Agency's expertise and technologies into inter-agency exercises directed at improving the protection and security of water systems and response to emergencies impacting water and air. These exercises and the subsequent partnership between NASA and PureSense helped to launch a unique solution for accelerating alert time and quickening the ability to respond.

In the spring of 2002, NASA and several of the eventual founders of PureSense sponsored a series of work sessions for emergency response and water system agencies, with representation from a wide variety of Federal, state, local, and first-response officials. The sessions focused on understanding how to deal more effectively with bioterrorist attacks.

The would-be founders of PureSense brought experience in building and using multidisciplinary applications of real-time intelligence systems and communications technology; these were integrated into a proof-of-concept hardware and software system application to support collaborative surveillance, detection, response, and recovery. The system demonstrated several of NASA's remote-sensing, advanced data-analysis, simulation, and scientific data-visualization technologies and repurposed them alongside other technologies and applications to facilitate cohesive, rapid decision-making across the agencies.

These experiences, coupled with a dedication to prevent water and air contamination from harming people and infrastructure, ultimately led to the launch of the PureSense company. PureSense, with the help of the NASA Technology Partnerships Division, initially opened an office on the NASA Research Park campus at Ames Research Center. The company recently moved to Emeryville, California, and continues to collaborate with NASA in bringing the Agency's expertise in several areas to projects concerned with homeland security.

Today, PureSense is providing customers in the commercial and municipal markets with the first cost-effective online service to detect, know, and respond immediately to critical changes in the quality of water and air, and to do so in real time. PureSense hardware and related firmware first integrate diverse remote sensors, instruments, and remediation technologies, and then network them for continuous, standard data acquisition. Additionally, PureSense software applications perform data integration, validation, and analysis to provide customers with continuous intelligence on critical conditions over secure communication channels and via multiple types of communication devices. Included in the analytical data model are data-mining algorithms developed by the company's founders, in collaboration with scientists at NASA. These algorithms have been repurposed and integrated into PureSense's air and water system-monitoring software product, known as the PureSense Environmental Resource Management (ERM) Solution.

Partnership

The PureSense founders, including team members who at one time worked in Ames's Biomedical Visualization and Gravitational Research laboratories, worked to reinforce the NASA-PureSense partnership through the signing of two Space Act Agreements in 2003, designed to support the NASA mission to "protect our home planet." Specifically, these agreements were designed to support the evaluation of NASA data-mining algorithms for incorporation in the analytics engine that is a component of the PureSense ERM Solution. These algorithms help NASA remotely monitor the physiological health of astronauts, evaluate atmospheric changes on Earth, conduct aviation system analysis, and discover and understand new galaxies.

The Space Act Agreements also led to PureSense partnering with the Agency's scientists on emergency-response protocols, software systems, and advanced data analysis. Equally important, this type of partnering helps PureSense to generate recognition in the environmental industry for relevant NASA research and development capabilities while also supporting NASA in its relationships with other government agencies, national laboratories, and key industry players concerned with the Nation's water and air quality. Subsequently, in 2004, PureSense signed a Cooperative Agreement with Ames to collaborate on U.S. Department of Homeland Security projects designed to ensure the security of the Nation's water systems. PureSense has also collaborated with NASA on projects to develop and improve techniques and tools used in urban search and rescue.

Product Outcome

Industry and the public are facing a rising tide of costs associated with the contamination of air and water. For example, in 2004, Chiron Corporation was unable to meet its flu vaccine production commitments due to microbial cross-contamination in a manufacturing facility. After an estimated $300 million loss in enterprise value and 6 months of plant downtime, the company regained its license to produce flu vaccine for 2005. Also, beverage companies have lost millions of dollars annually worldwide, as contamination during production has led to product recalls (e.g., The Coca-Cola Company's Dasani purified water was recalled after levels of bromate were found to exceed legal levels); lost inventory; and damaged brand identity. In 1990, North Carolina regulators issued a report stating that Perrier's bottled water was contaminated with benzene, a poisonous liquid shown to cause cancer in laboratory animals. Perrier attributed the contamination to an isolated incident, in which workers failed to filter out the benzene—which the company said is naturally present in carbon dioxide—during the filtration process.)

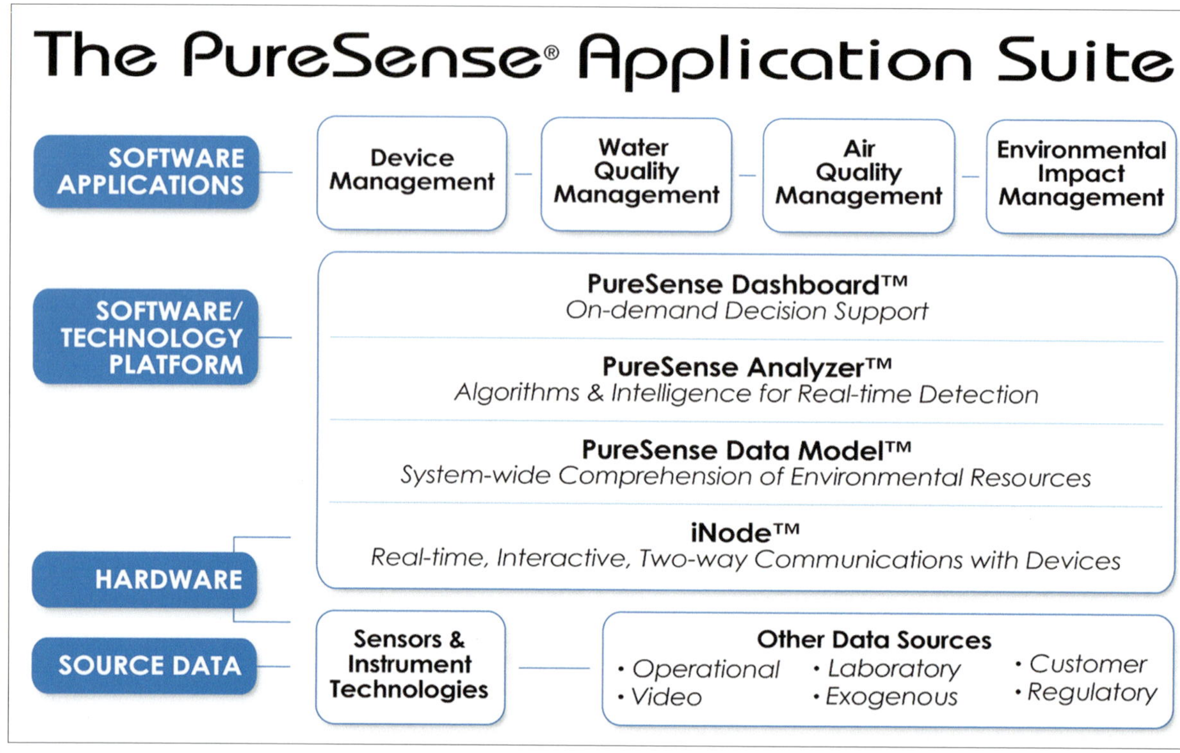

The PureSense Application Suite integrates four core product components (middle), fed by multiple data sources, to provide an information and communication platform, which produces comprehensive, real-time analysis of water and air quality for rapid detection, predictive analysis, and focused action.

In addition, U.S. hospitals incur $5 billion per year in direct costs from the transmission of infections inside their facilities that cause patients to become ill. Moreover, "sick building syndrome" in office buildings produces $30 billion to $50 billion per year in related health care costs and lost employee productivity in America. The State of California estimates that poor indoor air quality in non-industrial settings costs the economy $45 billion a year. Further, annual costs of asthma in the United States are estimated to be $18 billion, with $10 billion in direct costs, and another $3 billion in lost productivity from 15 million missed or lost work days. These experiences are more and more common as current practices in environmental monitoring fail to prevent problems from occurring.

PureSense's cost-effective ERM systems are shaping up to be the answer to the health and monetary issues affecting both industry and the public. PureSense's NASA-inspired solution helps transform the management of water and air from periodic batch processing to continuous, real-time understanding, based on automated, system-wide monitoring of environmental conditions.

The PureSense ERM Solution consists of a unique combination of hardware and software applications (known as the PureSense Application Suite) with the ability to access and aggregate data from multiple sensors and other data sources, add value to the data through analysis, and supply on-demand intelligence to users anytime, anywhere.

Four core components are fed by multiple data sources to provide an information and communications platform and produce comprehensive, real-time analysis of water and air quality for rapid detection, predictive analysis, and focused action.

The first component is a "pervasive computing" device, called the PureSense iNode, which provides two-way communication with existing or future air or water sensors. The iNode pre-processes information and transmits it over highly secured wireless connections to a centralized data center. The second component is a real-time data-collection network called the PureSense Data Model that, according to the company, gathers data faster and more economically than current "grab sample" methods (involving manual collection of samples for laboratory analysis) and enables the integration of disparate data sets from multiple sources. The third component, the PureSense Analyzer, employs sophisticated algorithms and NASA-influenced data-mining techniques to correlate multiple streams of data to produce real-time alerts and other analyses regarding environmental conditions. The fourth component, the PureSense Dashboard, is tailored to individual user profiles and communicates critical, real-time analysis to users in a timely manner. In the water industry, this is the powerful, role-based response system that keeps all responsible parties informed about key changes in water quality.

PureSense Environmental, Inc., has collaborated with NASA on projects to develop and improve techniques and tools used in urban search and rescue. To the left, first-response officials work together during a collapsed building exercise.

PureSense has developed aggressively over the last year, creating customer traction and new projects on the national level, with the U.S. Environmental Protection Agency and the Department of Homeland Security, among others. The company chose municipal drinking water as the best avenue for commercialization of existing NASA technologies and has already completed initial customer installations. Early endorsements for the PureSense ERM Solution have been received from some of the Nation's largest water companies, including the Los Angeles Department of Water and Power, the second-largest municipal water company in the United States, and several other water companies that each serve over a million customers, including the East Bay Municipal Utility District, the Las Vegas Valley Water District, and the San Jose Water Company.

PureSense is currently expanding the capabilities of the PureSense ERM Solution for faster analysis and remediation in new vertical markets. The initial expansion includes hospitals and biopharmaceutical manufacturing, where it can help mitigate infectious disease incidences and, hence, drive down operational risks and associated costs. PureSense is now working with one of the largest hospital systems in California to adapt the PureSense ERM Solution for indoor air quality management. The company is also exploring interest from the food and beverage industries, where it can help minimize downtime and materials spoilage.

The partnership between NASA and PureSense has resulted in a business that is poised to have a dramatic impact on many different industries and on the lives of many people, with the potential to revolutionize environmental monitoring and control. ❖

PureSense® is a registered trademark of PureSense Environmental, Inc. Environmental Resource Management,™ PureSense iNode,™ PureSense Data Model,™ PureSense Analyzer,™ and PureSense Dashboard,™ are trademarks of PureSense Environmental, Inc.

Remote Sensing for Farmers and Flood Watching

Originating Technology/ NASA Contribution

The Applied Sciences Directorate, part of NASA's Science Mission Directorate, makes use of the Agency's remote-sensing capabilities to acquire detailed information about our home planet. It uses this information for a variety of purposes, ranging from increasing agricultural efficiency to protecting homeland security. Sensors fly over areas of interest to detect and record information that sometimes is not even visible from the ground with the human eye. Scientists analyze these data for a variety of purposes and make maps of the areas. These maps are often used to answer questions about the environment, weather, natural resources, community growth, and natural disasters.

Partnership

Located at Stennis Space Center, Mississippi, and founded in 2002, NVision, Inc., is a geospatial information systems company that has tapped into NASA's wealth of remote-sensing information. NVision is a small, minority, woman-owned business with two **Small Business Innovation Research (SBIR)** contracts under development at Stennis. Even though the research is still underway, several products related to the work have already come to market.

Product Outcome

NVision harnessed NASA's geospatial satellite information to provide innovative geospatial solutions for a variety of industries. It provides tailored solutions for customers' needs and, as a result, has made three rather disparate spinoffs: a crop prescription service for farmers; a disaster management tool for local, state, and Federal governments; and an educational service for young farmers.

The first is a service available to farmers and those in the agricultural community. NVision commercialized this system through InTime, Inc., a precision agricultural company providing farmers with automated, digital crop prescriptions within 24 hours of aerial data collection. InTime is another high-tech company that originated at Stennis. The service allows customers to generate their own prescriptions and crop scouting maps at any time of the day or night, using Web-based technologies built in collaboration with NVision that harness over 25 years of NASA precision agriculture algorithms and research. They can print scouting maps showing relative crop health as well as cost reports showing the economics of treating a field with herbicides, insecticides, plant growth regulators, and defoliants. These maps come in electronic formats and as hard copies. Growers can rapidly verify the scouting maps and then download the digital prescription, which is then loaded into an inexpensive Global Positioning System-enabled, hand-held computer on the farmer's sprayer equipment. The farmer loads the appropriate fertilizer or crop controller, and then treats the specific area.

The ability to focus treatment saves time and money, as manpower and product can be used efficiently. In

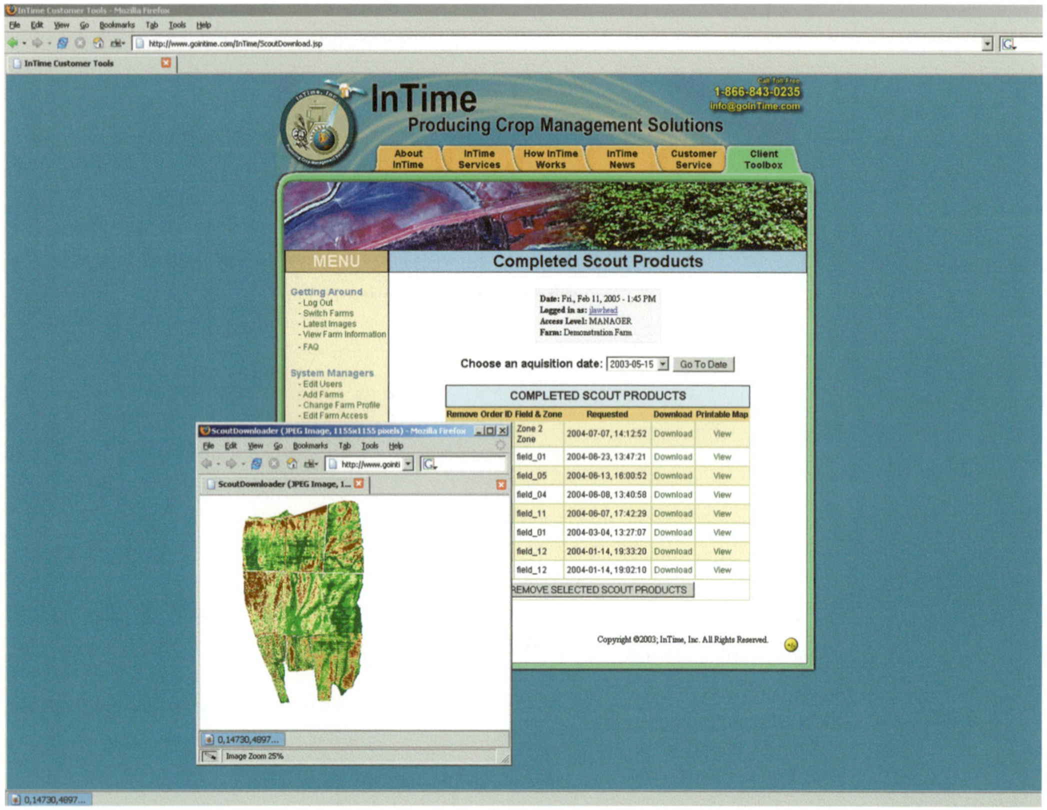

InTime, Inc., provides crop management services that target a reduction of chemical costs by providing farmers with prescription maps.

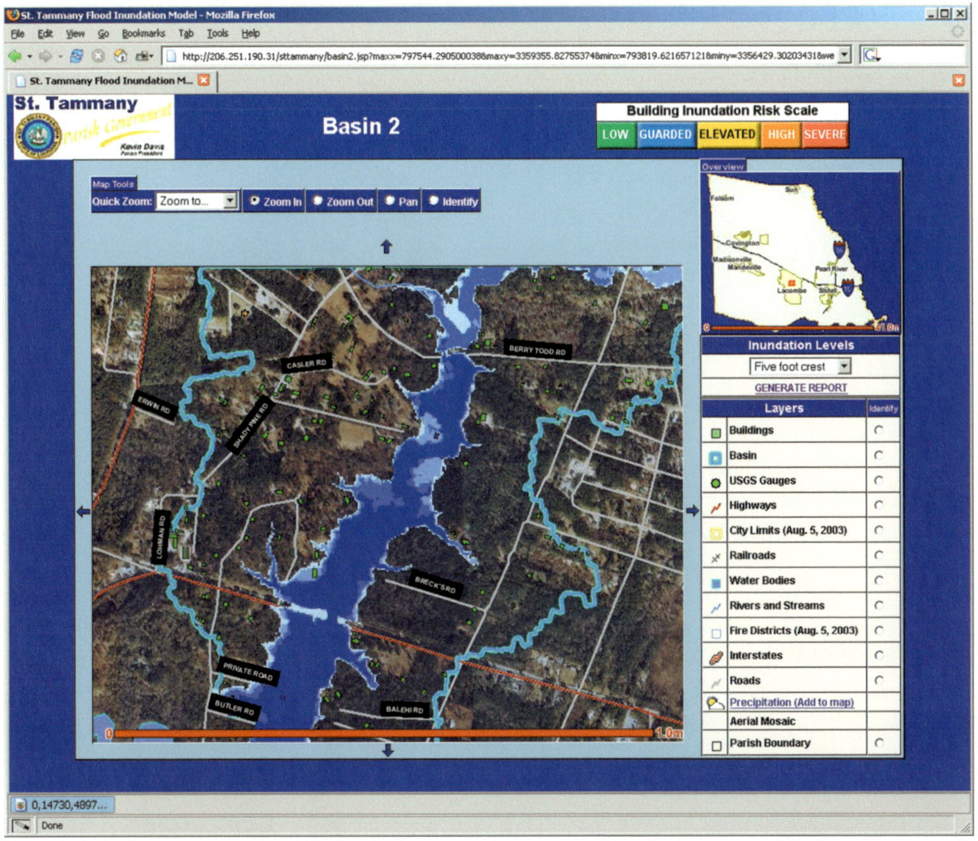

St. Tammany Parish, Louisiana, frequently under flood watch, has a new tool to help its citizens when the waters rise.

the case of inorganic pesticides and fertilizers, this approach provides an added environmental benefit, as chemicals with the potential of entering groundwater are used sparingly.

In addition to the work done with the SBIR awards, NVision also partnered with NASA and the government of St. Tammany Parish, Louisiana, to produce an emergency response decision management system through a dual-use contract with Stennis. St. Tammany Parish is in the southern Louisiana flood region and experiences sweeping flood waters unpredictably each flood season.

The local government, NASA, and NVision teamed up to create the Real-time Emergency Action Coordination Tool (REACT). REACT is a simple, Web-based tool that city officials can access when they need to make decisions in emergency and disaster situations. It provides a comprehensive network of maps and reports, combined with real-time sensors, shelter and hospital information, and dynamically generated environmental model output during a crisis to help officials make timely, informed decisions under pressure.

While the system cannot prevent flood waters from rising, it does provide citizenry with up-to-date information about where the water will be next and where it is safest for them to be. The local government can have emergency dispatchers alert residents and warn them of the danger, which could save lives and thousands of dollars in damage. Use of the system is not restricted to floods, but can be applied to virtually any type of disaster, including terror attacks, fires, and hazardous material spills.

REACT was successful in St. Tammany Parish and NVision has been contracted to create two additional prototypes. It is currently under contract to install a REACT system in nearby Jefferson Parish, Louisiana. A third system is under negotiation to be installed in Medford County, New Jersey.

The collaboration between NVision, NASA, and the local governments has been so strong that NVision won the Louisiana Governor's "Technology Innovation of the Year" award for 2004 and the Mississippi "Small Business Innovator" award for 2005. Additionally, NVision was named an Environmental Systems Research Institute New Business Partner of the Year for 2005.

NVision has also worked with NASA's Ag 20/20 program at the 2002 Farm Progress show in Alleman, Iowa, where it received a warm response from the youth who were fascinated by NASA's high-tech approach to farming. This experience prompted a partnership between NASA's Agricultural Science Division, NVision, and the Future Farmers of America. The three worked together to create an educational, geospatial-based, precision agriculture application to distribute free of charge to students nationwide to familiarize the next generation of farmers with geospatial technology and to encourage them to take full advantage of NASA science. More than 1,000 copies of the software were distributed via the Internet to youth in 30 states and 9 countries. ❖

Pesticide-Free Device a Fatal Attraction for Mosquitoes

Originating Technology/ NASA Contribution

Are those pesky mosquitoes getting more entertainment out of your family picnic than you are? If the answer is yes, then it is time to reclaim your backyard with assistance from an unlikely partner.

Nowadays, NASA is developing tools to track and predict the spread of the West Nile Virus on a global scale, but several years ago, the Space Agency carved out some time to collaborate with an outdoor products manufacturer in order to help control mosquito populations on a local level. The technology resulting from this union leveraged a space-age heat blanket to attract mosquitoes, which would then be eliminated without the use of harmful pesticides or chemicals.

Partnership

In 1995, Alvin Wilbanks, of Environmental Products and Research, Inc., met with Marshall Space Flight Center's Technology Transfer Office to discuss his ideas of developing a mosquito-killing system. Wilbanks's proposal for such a product evolved from nightly rituals of coming home from work and having to free his house of mosquitoes in order to protect his newborn daughter. He had been applying repellents to her, but was apprehensive about what could happen if she accidentally ingested them or got them in her eyes.

Prior to meeting with NASA, Wilbanks started looking into pesticide-free products, only to find that no such items were available. He knew there had to be a better way to control mosquitoes without the use of chemicals, so he asked himself, "What attracts the mosquito to us?"

His first theory was that the mosquito was seeking blood; he placed blood on a slide as an experiment to view the mosquito's behavior. To his astonishment, the mosquito was not attracted to the blood sample. He then began to test other methods of attraction, such as moisture, body scents/pheromones, motion, and breathing. All of these experiments had poor results, as there were no increases in the ratio of capturing and killing the insect. Wilbanks even tried an off-the-shelf bug light, but after examining the contents, it was evident that the mosquitoes were not being drawn by this method. Without giving up on the bug light unit, Wilbanks removed the lights and replaced them with a heating element. The test revealed that most species of mosquitoes were attracted to the heat emitting from the modified unit, yet they were not being captured.

He spent the next 3 years determining what temperature would be most effective and how big the unit should be in order to achieve the most efficient results. Other tests were performed to determine effectiveness, including air flow and wind design. The final analysis indicated that temperature, size, motion, and elevation were all determining factors in capturing and killing mosquitoes. Armed with these facts, Wilbanks was closing in on the development of his own mosquito-killing system.

There was a delay in the process, however, as the mosquitoes were smartening up and detecting the electrical fields of Wilbanks's experimental killing system. It turned out that the electrical fields were acting as warning signals to the insects, thus deterring them from investigating further. Wilbanks decided to alter the unit in order to conceal the electrical fields but still attract and kill the mosquitoes. This added other variables that had to be perfected before the system was ready to go to market. It was at this point when Wilbanks submitted a request for technical assistance to NASA's Marshall facility.

Marshall engineers suggested a number of improvements to cut costs and enhance the efficiency of Wilbanks's system. The specific areas of assistance included the difficulties identified with the electrical fields and issues pertaining to the heat source. Based on these suggestions, Wilbanks decided to use a NASA-influenced heat blanket in his system. This technology simulates body heat and skin to attract the mosquitoes. Additionally, Marshall referred Wilbanks to a nonprofit agency that aided him

The Mosquito Killing System was developed with technical assistance from NASA and is an environmentally safe way to reduce the mosquito population.

in developing the prototype mosquito-killing system as well as the molds used to form the base and top structure of the product.

After experimenting on 40 different versions, Wilbanks came up with a mosquito-killing device that met his satisfaction. He was granted a patent for his innovation in 1997, followed by Underwriters Laboratories (UL) approval in 1999.

Product Outcome

Now owned by Arctic Products, of Jefferson City, Missouri, the Mosquito Killing System utilizes the unique heating source inspired by NASA, along with carbon dioxide, to lure mosquitoes within a 1-acre radius. Since mosquitoes' natural "hunting" techniques involve heat sensing and carbon dioxide detection, the Mosquito Killing System employs heat and the carbon dioxide to mimic body temperatures and breathing patterns of humans, livestock, poultry, and domestic pets. Every 6 to 9 minutes, the killing system cycles through various temperatures and releases carbon dioxide to attract the mosquito.

Once inside the unit, the mosquito is forced by a vacuum-powered fan into a grid that provides an electrical charge to safely eliminate the insect. The remains are then expelled through the bottom of the unit where they can either be collected by an attached mesh bag for disposal or returned naturally to the environment. The Mosquito Killing System is equipped with an energy efficient photocell that turns the unit on at dusk, and off at dawn. (Estimated operational cost is $6 to $8 a month.)

Unlike products that use propane or chemicals, the Mosquito Killing System is as environmentally friendly as having an additional tree in the yard, since it stores carbon dioxide and releases it back into the atmosphere. The system uses a standard 20-pound carbon dioxide tank—the same tank used throughout the beverage industry. Carbon dioxide provides a safe alternative to chemicals and pesticides found in competing products, because it is nonflammable, nonexplosive, and nontoxic.

The system captures several times as many mosquitoes as any other machine or product, according to the manufacturer. The female mosquito—which feeds on blood from people, animals, and birds, unlike its male counterpart—lays an average of 300 eggs at a time. In essence, for every 1,000 female mosquitoes eliminated, the mosquito population is actually reduced by 300,000.

Not only can consumers diminish the local mosquito population with the Mosquito Killing System, they can reduce the risk of various diseases vectored by mosquitoes and provide for a cleaner environment. In addition, the system will attract and eradicate other pests such as gnats, biting flies, and "no-see-ums," but unlike conventional bug "zappers," it will not draw the attention of beneficial insects that are good for the environment, such as pollinators and species that feed on plant-eating pests to help preserve gardens.

From backyard barbecues and ballparks, to stables and campground get-togethers, the Mosquito Killing System is ideal for just about any outdoor application. The Belmont Park Racetrack, in Elmont, New York, for instance, is using the technology to stave off mosquitoes from its prize-winning thoroughbred horses. "There is no acceptable risk when it comes to introducing chemicals around thoroughbreds," claimed John Chilcott, the general foreman of Belmont Park. "We could use anything to control mosquitoes, and we chose the [Mosquito Killing System]. It is environmentally safe and we are completely satisfied." ❖

Mosquito Killing System™ is a trademark of Arctic Products.

How the Mosquito Killing System works: 1) A photocell activates the unit at dusk and powers off at dawn. 2) A heat source mimics the body temperatures of humans, livestock, poultry, and domestic pets. 3) An automated carbon dioxide release mimics the breathing of warm-blooded animals. 4) A fan provides a vacuum source to force the mosquito through the unit. 5) An electronic grid provides an electrical charge to safely eliminate the mosquito.

Making the Most of Waste Energy

Originating Technology/ NASA Contribution

The Thermo-Mechanical Systems Branch at NASA's Glenn Research Center is responsible for planning and conducting research efforts to advance thermal systems for space, aerospace, and non-aerospace applications. Technological areas pertain to solar and thermal energy conversion. For example, thermo-mechanical systems researchers work with gas (Stirling) and liquid/vapor (Rankine) systems that convert thermal energy to electrical power, as well as solar dynamic power systems that concentrate sunlight to electrical power.

The branch's development of new solar and thermal energy technologies is propelling NASA's missions deep into unfamiliar territories of space. Solar dynamic power systems are actively improving the health of orbiting satellites, giving them longer life and a stronger radiation tolerance, thus, creating less need for on-orbit maintenance. For future missions, NASA may probe even deeper into the mysterious cosmos, with the adoption of highly efficient thermal energy converters that have the potential to serve as the source of onboard electrical power for satellites and spacecraft. Research indicates that these thermal converters can deliver up to 5 times as much power as radioisotope thermoelectric generators in use today, for the same amount of radioisotope.

On Earth, energy-converting technologies associated with NASA's Thermo-Mechanical Systems Branch are being used to recover and transform low-temperature waste heat into usable electric power, with a helping hand from NASA.

Partnership

In 2003, Mount Prospect, Illinois-based Unitel Technologies, Inc., approached NASA with an idea for an advanced energy recovery cycle that it believed could cost-efficiently convert low-level thermal energy sources from previously untapped resources—such as hot gas exhausted from power plants—into usable electric power.

According to Unitel Technologies, each and every day, all around the world, an incalculable amount of energy is wasted and literally blown into the atmosphere through power plant smokestacks and industrial and commercial heating systems. Billions of energy units are additionally carried away by the cooling water and air that are used in many of the related processes, the company added.

"Power plants, be they stationary or vehicle, are typically not very good at squeezing the last few BTUs (British thermal units—used to measure heat created by the burning of any material) out of their fuel source," noted Serge Randhava, president of Unitel Technologies. "Much of this lost potential is simply blown up the stack or out of exhaust pipes, in the form of low-level waste heat. The volumes of this thermal energy are unfathomably large, but the temperatures are low, and conversion of low-level heat sources into usable electric power can be difficult," he added.

Harnessing the Sun's energy through solar thermal propulsion could propel vehicles through space by significantly reducing weight, complexity, and cost while boosting performance over current conventional upper stages. Pictured is an artist's conception of space flight using solar thermal propulsion.

The company's proposal to NASA aimed to address this issue. As it turned out, however, the Space Agency already possessed some expertise in this area. Researchers from Glenn's Thermo-Mechanical Systems Branch developed thermodynamic-analysis software to aid in the recovery of the exhaust heat, or waste heat, from the Rankine-cycle engines of M1 Abrams military battle tanks. The two parties agreed to enter into a partnership in which their knowledge would be shared to advance thermal energy recovery efforts throughout industry.

Unitel Technologies received a NASA grant to design a prototype waste heat recovery system called NEOGEN. The goal was to achieve a nominal energy-savings gain of approximately 20 percent over the recuperated binary Rankine cycle. The work was carried out with assistance from the Glenn researchers and engineers from Creare, Inc., who aided in the development of the thermodynamic-analysis software. Moreover, an award from the NASA Illinois Commercialization Center supported NASA's thermodynamic-efficiency analyses of NEOGEN.

With regard to the prototype's design, the idea was to create a system that could use a unique absorption cycle to tap into waste heat streams of 125 °C (257 °F) to 400 °C (752 °F). Using the thermodynamic-analysis software, Glenn researchers modeled the cycle efficiencies, optimized the heat exchanger design, and provided operating cost data. In addition, NEOGEN performance was benchmarked against related systems that NASA worked with in the past.

The results of this cooperative research effort were successful. The three parties took a concept for a new heat recovery cycle, with a unique and flexible working solution, all the way to demonstration-ready design.

"Working with NASA let us dramatically short-circuit our development program," said Randhava. "We were able

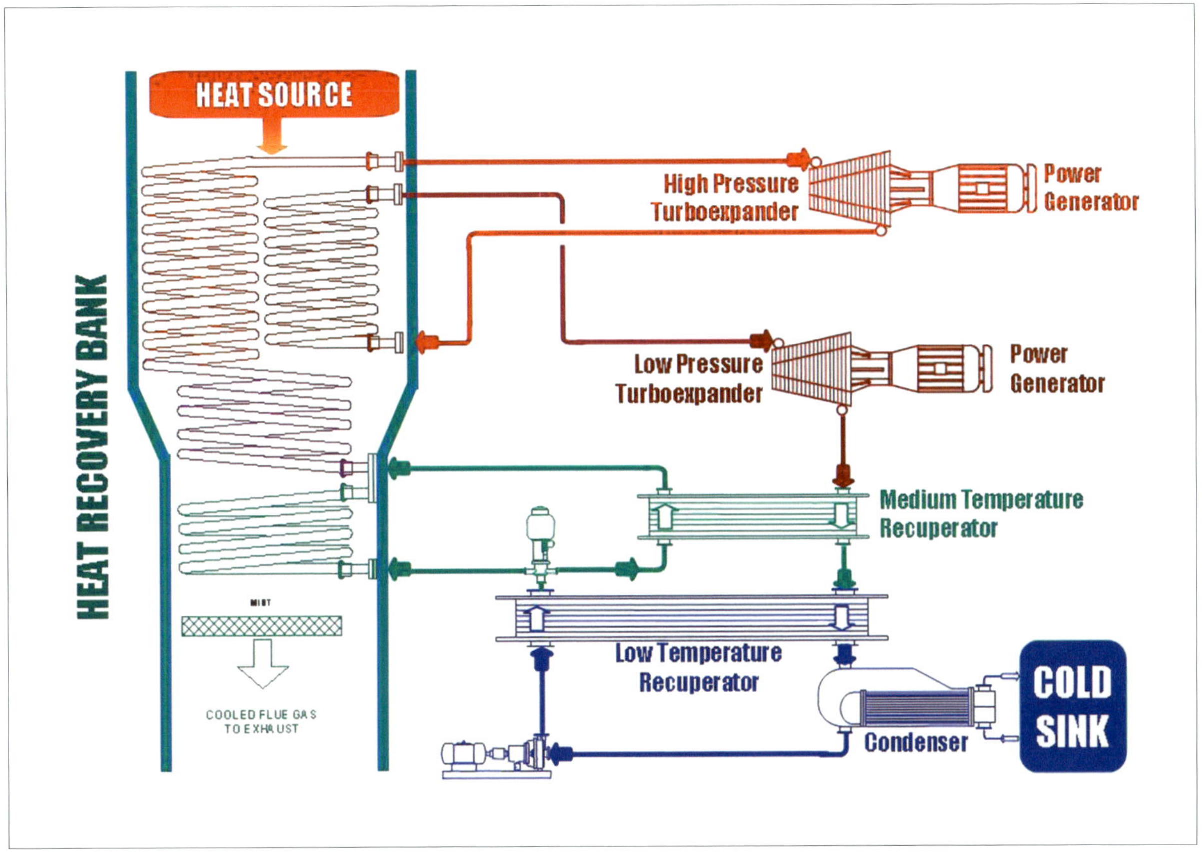

This system schematic was developed by Unitel Technologies, Inc., Creare, Inc., and Glenn Research Center. The ensuing waste heat recovery unit features a twin-turbo generator and a twin-recuperator configuration that give it great flexibility in heat-source harvesting.

to go directly from concept to prototype design phase without the need for endless bench and pilot scale experiments."

On top of the time savings, the partnership with Glenn saved Unitel Technologies over $250,000 in developmental costs.

Product Outcome

While a heat recovery system able to operate in the lower spectrum of waste temperatures opens up an enormous number of applications for exploitation, Unitel, with continued guidance and resource-support from NASA, decided to tiptoe into commercialization with a single focus on marine power plants, where everyday, just one single large ship can generate billions of BTUs in the form of low-level waste heat.

According to the company, marine power plants have a few unique characteristics that make them well-suited for waste heat recovery, and a few which make them a challenge. One advantage is that there are vast, cold sinks sloshing beneath every marine power plant outside of dry dock. Unitel, NASA, and Creare designed the NEOGEN system based on a 27 °C (81 °F) cold sink—equivalent to oceanic temperatures. On the challenging side, there is limited space for a recovery unit to operate within the power plant of a marine vessel, plus, there are many safety considerations that come with operating such machinery on a ship.

"If the NEOGEN system can help us harvest even a modest fraction of this energy stream, we can save a lot of fuel and help our environment at the same time," claimed Randhava.

Beyond the initial marine power plant applications, the company plans to use its system to recover energy from other industrial plants, as well as commercial plants, electric utility facilities, diesel engines, and gas turbines. The company is looking to gain a foothold in these areas that it said were largely disregarded until about 4 years ago, when the price of crude oil started its upward climb.

"Most of the waste heat recovery strategies in use today are directed at high-temperature streams, usually over 400 °C (752 °F). The low-temperature spectrum has been ignored because the driving economic pressures were simply not there," said Randhava. "The decision has now become highly significant, given the fact that the price of crude oil is more than $60 per barrel. Every BTU of energy that can be saved means money in the bank, more so now than ever." ❖

Washing Away the Worries About Germs

Originating Technology/ NASA Contribution

Fresh fruits and vegetables have been in demand by orbiting astronauts since the early days of the Space Shuttle. As one can imagine, however, oranges, onions, tomatoes, garlic, and other fresh items can provide a cornucopia of smells in a closed environment such as the Space Shuttle or the International Space Station (ISS), especially when they begin to perish. It does not help that they are loaded onto the Space Shuttle up to 24 hours in advance of a launch, and that the on-orbit shelf life is just 2 to 3 days for most, due to a lack of refrigeration.

While such produce adds significant variety to astronauts' diets and increases their morale, the odor that emanates from it as it ages can cause nausea. One of the last things astronauts need is associating this healthy fare with feelings of nausea.

NASA is currently investigating the use of a commercial sanitation product it helped develop with private industry to thoroughly cleanse and, thus, increase the shelf life of fruits and vegetables being sent into space. Meanwhile, this product is ripe for the picking for consumers looking to do the same, and more, on Earth.

Partnership

Microcide, Inc., was incorporated in 1990, with the primary objective to develop unique, non-toxic, and environmentally safe microbicidal products for personal care, public health, food, and agriculture, through research and development. The Troy, Michigan-based company was introduced to NASA through the Glenn Research Center-sponsored Garrett Morgan Commercialization Initiative (GMCI), a program for small, minority- and woman-owned businesses that can benefit from NASA resources.

As a result of this meeting, Microcide was welcomed to join the NASA Food Technology for Commercial Space Center (FTCSC) as an affiliate partner, with sponsorship from GMCI. The mission of the NASA FTCSC is to lead a national effort in developing foods and food-processing technologies that enhance the safety of space missions and advance commercial food products through cooperative efforts with NASA scientists and technologists, commercial companies, and academic researchers.

As an affiliate partner, the company interacted with personnel from Johnson Space Center and learned that there was a need for a nontoxic, biodegradable, microbicidal product to disinfect fresh fruit and vegetables for Space Shuttle crews. Hydrogen peroxide and hypochlorite are the oxidizing agents that have been used as sanitizing agents in past space missions, but the efficacy of hydrogen peroxide remains questionable and both hydrogen peroxide and hypochlorite are known to produce carcinogenic free radicals, according to published research.

Microcide developed PRO-SAN, a technology comprised of safe sanitizing agents that could possibly be used as an alternative to the two controversial oxidizing agents.

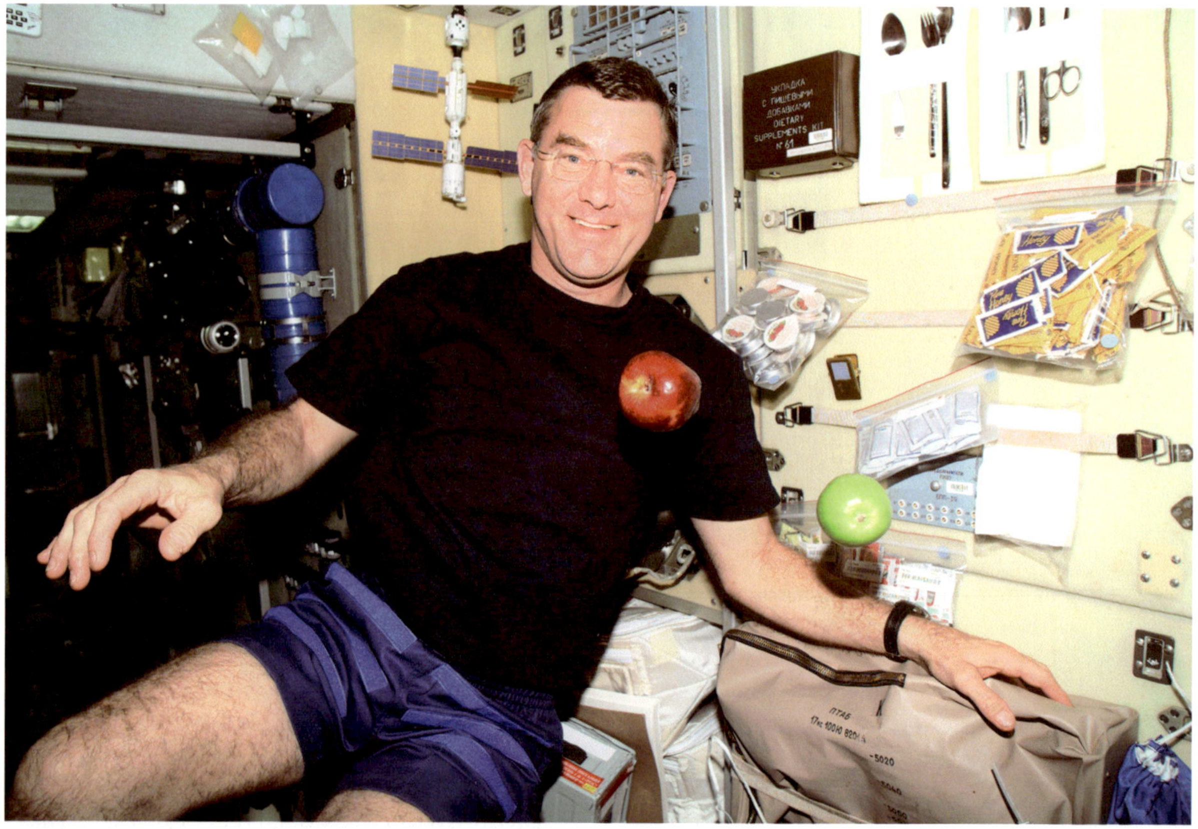

Astronaut James Voss, Expedition Two flight engineer, is deciding between two colors of apples as he takes a break for a snack in the Zvezda Service Module on the International Space Station.

Dr. Anthony Pometto, director of the NASA FTCSC, and Vicki Kloeris, manager of Space Shuttle and ISS food systems at Johnson, wondered whether PRO-SAN could improve current food-sanitizing measures in space, so they commissioned Dr. Aubrey Mendonca, an FTCSC outreach mission specialist, and Dr. Cheryll Reitmeier, an FTCSC education mission specialist, to carry out a comparative investigation of the sanitizer against various other disinfecting products.

The study proved the potential of PRO-SAN as a safe, stable, and biodegradable product for use in space; however, at the time, it was only available as a concentrated powder and a ready-to-use liquid. The powder is difficult to handle in zero gravity and the use of liquid presents increased payloads for flights. NASA and Microcide decided that the alternative would be to have the PRO-SAN powder concentrate available as a water-soluble package. Once this package is dropped in water, it can be stirred until it dissolves, creating a ready-to-use sanitizer.

Based on this research, Microcide has perfected a food-grade soluble packaging design for use on Earth.

PRO-SAN sanitizing products offer safety from harmful foodborne bacteria, such as *E. coli*, *salmonella*, *listeria*, *cholera*, *shigella*, *staphylococcus*, and *streptococcus*.

Produce Outcome

The water-soluble powder concentrate formula of PRO-SAN is now available, in addition to the ready-made spray bottle formula. The difference between the two is that the water-soluble version is 100 times more concentrated than the spray version.

Microcide asserts that all PRO-SAN products offer safety from harmful foodborne bacteria, such as *E. coli*, *salmonella*, *listeria*, *cholera*, *shigella*, *staphylococcus*, *streptococcus*, and many others. PRO-SAN even instantly kills antibiotic-resistant bacteria, the company adds.

The sanitizer is free of volatile organic compounds, chemicals that have been found to be a major contributing factor to ozone pollution. Simply stated, PRO-SAN does not pollute indoor air in enclosed spaces of homes and kitchens. Also important is that the food-grade ingredients and emulsifying agent of PRO-SAN do not produce carcinogenic by-products or free radicals.

The technology has three different functional properties of cleaning, sanitizing, and chelating (a process that uses chemical substances to bind molecules and, thus, remove toxins), with the power to kill 100 million bacteria in 30 seconds, with 99.999-percent efficiency, according to tests. PRO-SAN-treated fruits and vegetables maintain their original taste and flavor, not to mention that even their natural colors are enhanced—all without leaving behind an unwanted film or residue. In addition to fruits and vegetables, PRO-SAN can be used on counter tops, cutting boards, dishes, utensils, forks, knives, and sinks.

Ideal environments for the use of PRO-SAN include hospitals, nursing homes, day care centers, dormitories, restaurants, and offices.

In 2004, Microcide was awarded a Phase I **Small Business Innovation Research (SBIR)** contract with Johnson to further explore how nontoxic sanitizing methods can prevent food contamination and improve hygienic conditions during prolonged space missions and here on Earth. ❖

PRO-SAN® is a registered trademark of Microcide, Inc.

Celestial Software Scratches More Than the Surface

Originating Technology/ NASA Contribution

While NASA is preparing to send humans back to the Moon by 2020 and then eventually to Mars, the average person can explore the landscapes of these celestial bodies much sooner, without the risk and training—and without even leaving the comfort of home.

Geological data and imagery collected from NASA missions are enabling anybody with computer access to virtually follow the footsteps of Apollo astronauts who walked on the Moon or trace the tracks of the exploration rovers currently on Mars.

Partnership

Reading Information Technology, Inc. (RITI), of Reading, Massachusetts, received access to Moon- and Mars-mapping data from NASA via the U.S. Geological Survey (USGS), a Federal source for reliable, scientific information. The technical service and software-engineering company adapted the NASA data to create a powerful software suite that lets users explore accurate 3-D models of the Moon and Mars.

In developing its lunar-mapping software, RITI configured Moon data retrieved from the Deep Space Program Science Experiment. Commonly referred to as the "Clementine" mission, this experiment was a joint project between NASA, the Ballistic Missile Defense Organization (now the Missile Defense Agency), the U.S. Naval Research Laboratory, and the Lawrence Livermore National Laboratory. The objectives were to test sensors and spacecraft components under extended exposure to the space environment and observe both the Moon and the near-Earth asteroid, 1620 Geographos. (The Geographos observations were not actually made due to a malfunction in a thruster that caused the spacecraft to spin uncontrollably.)

The Moon observations, made at various wavelengths in the visible spectrum, as well as in ultraviolet and infrared, were processed (post-mission) by the USGS for the purposes of obtaining multispectral imaging of the entire lunar surface and assessing the surface mineralogy. NASA's Goddard Space Flight Center captured and processed the altimetry and gravity data.

The Clementine spacecraft that collected all of this information was launched on January 25, 1994, from Vandenberg Air Force Base, California. It was the first of a new class of small spacecraft to enable long-duration deep space missions at low cost, using lightweight satellite technology.

After two Earth flybys, lunar insertion was achieved on February 19. Mapping of the Moon's 38 million square kilometers took place over approximately 2 months, in two systematic mapping passes of the orbiting body. Four cameras were used to take over 2 million images. Roughly 50,000 of the images were processed and mosaicked to produce a global map of the Moon's albedo (normalized brightness or reflectivity). The mission resulted in the first ever uniform global lunar topographic map. Additionally, the Pentagon announced in 1996 that the data acquired by the Clementine spacecraft indicate that there is ice located at the south pole of the Moon. Deposits of ice on the Moon could have many practical uses for future manned exploration, because lunar water could serve as a source of oxygen, as well as hydrogen, which could be used as rocket fuel.

Following on the lunar-mapping software, RITI released a Mars-mapping program that is based on data tapped from NASA's ongoing Mars Exploration Rover mission, as well as information from previous assignments surveying the Red Planet, such as the Viking, Pathfinder, and Global Surveyor missions. These endeavors returned high-resolution images of volcanoes, lava plains, canyons, craters, wind-formed features, and areas where water once flowed, helping NASA to characterize the structure and composition of the planet's atmosphere and surface.

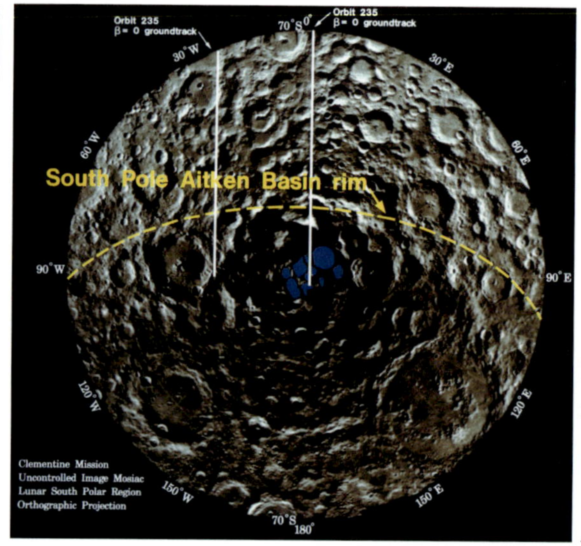

Image courtesy of the U.S. Naval Research Laboratory.

Data acquired by the Clementine spacecraft indicate that there is ice in the bottom of a crater on the Moon.

RITI continues to draw from the current rover mission, as there is still much to be learned about Earth's nearest neighboring planet.

Product Outcome

In most cases, the expression "ask for the Moon" means to ask for something that is typically out of reach or just about impossible to obtain; but in the case of RITI's lunar-mapping software, ask and you shall receive.

Advertised by RITI as the "most advanced, most detailed, and highest resolution lunar software on the market," Lunar Map Pro is a complete and comprehensive mapping program for astronomers, astronomy students, and casual Moon observers. The maps are corrected in real time with full libration (tilting) so that users can accurately visualize the different phases of the Moon. They can also be customized to match the flipped and reversed views experienced through users' telescopes and eyepieces.

Lunar Map Pro hosts a database of approximately 8,200 near-side and far-side features, plus an almanac that provides real-time data for observing locations. Geographic information system navigational tools and advanced search capabilities let users identify any physical feature of the Moon with pinpoint accuracy at the click of the mouse; users can also measure the actual sizes of and distances between features and keep observational notes.

Lunar Map Pro continues to transport users well beyond the Moon's surface with a 3-D viewer. This function allows the individuals to create precise 3-D displays of the rugged landscape, based on actual digital elevation models supplied by NASA. Additionally, a resource viewer provides a variety of animations and video clips, as well as easy access to numerous lunar Web sites.

A virtual walk on the Moon not enough? For those having a mania for Mars, it seems that the best way to get there without going there is to strap into your desk chairs and load RITI's Celestial Explorer: Mars into your computers.

According to RITI, Celestial Explorer: Mars contains the most complete Mars-feature database commercially available. Possessing features and tools similar to its Moon-mapping sister product, Celestial Explorer: Mars takes users on a voyage around the Red Planet, allowing them to explore and measure actual canyons, mountains, and craters—all of which can be called up by name in order to acquire historical facts and geological information. Physical elements of interest include the Vallis Marineris, a vast canyon system that is as long as the United States and 6 to 7 times deeper than the Grand Canyon, and the Olympus Mons, the largest known volcano in the solar system. These geographical wonders, and more, can be displayed in 3-D for the ultimate realistic experience.

As the Mars Exploration Rover mission carries on, RITI continues to update its Mars package with the latest video and 360-degree images taken from the Spirit and Opportunity robotic explorers, as well animated clips and educational games for students.

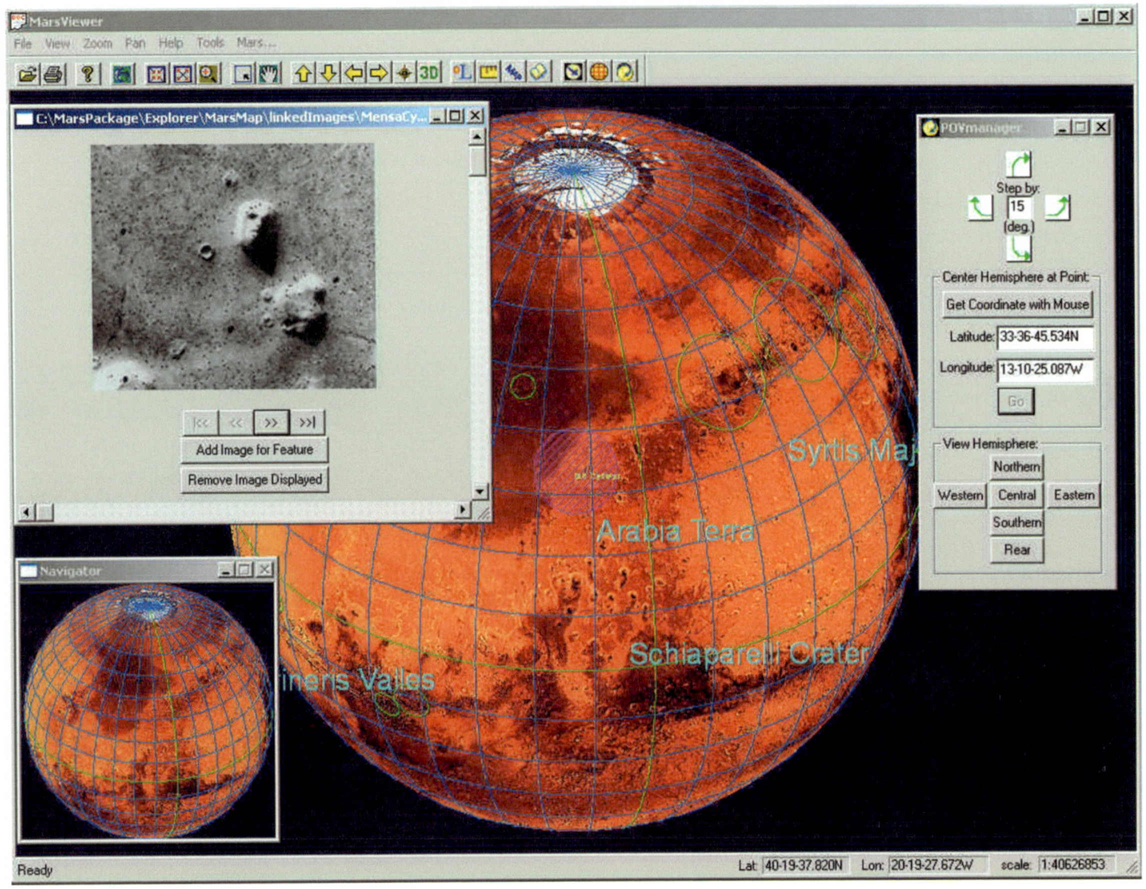

Celestial Explorer: Mars comes complete with high-definition Mars maps from NASA. Using these maps, individuals can plot coordinates to precisely locate areas of interest, such as the planet's Cydonia Mensae region. In 1976, the Viking 1 spacecraft photographed a rock formation in this region that resembled a human face (top left corner). High-resolution photos taken since indicate that the formations seen in the original, low-resolution image likely resulted from unusual lighting and shadow conditions, as they now look much less distinct and more like a natural feature of the landscape.

A diverse pool of users are benefiting from Lunar Map Pro and Celestial Explorer: Mars, including NASA's Johnson Space Center and celebrated science fiction writer Sir Arthur C. Clarke.

The two products are also receiving high marks from industry-related publications. For example, Sky and Telescope magazine recently called Lunar Map Pro "the most detailed software available" and Celestial Explorer: Mars "the next best thing to being [on Mars]." ❖

Lunar Map Pro™ and Celestial Explorer: Mars™ are trademarks of Reading Information Technology, Inc.

A Search Engine That's Aware of Your Needs

Originating Technology/ NASA Contribution

Internet research can be compared to trying to drink from a firehose. Such a wealth of information is available that even the simplest inquiry can sometimes generate tens of thousands of leads, more information than most people can handle, and more burdensome than most can endure.

Like everyone else, NASA scientists rely on the Internet as a primary search tool. Unlike the average user, though, NASA scientists perform some pretty sophisticated, involved research. To help manage the Internet and to allow researchers at NASA to gain better, more efficient access to the wealth of information, the Agency needed a search tool that was more refined and intelligent than the typical search engine.

Partnership

NASA funded Stottler Henke, Inc., of San Mateo, California, a cutting-edge software company, with a **Small Business Innovation Research (SBIR)** contract to develop the Aware software for searching through the vast stores of knowledge quickly and efficiently. The partnership was through NASA's Ames Research Center.

Product Outcome

Stottler Henke's Aware software helps researchers deal with the mass of information that comes from tapping into the Internet. With user-focused search technologies, it weeds through throngs of documents and produces personalized, focused results. By providing researchers with tools that augment their decision-making abilities, Aware allows them to quickly access and digest the wide range of information pertinent to their tasks.

Currently, Internet users have at least four ways of accessing information they need. First, people can query any one of a number of search engines, such as AltaVista. Second, users can search through existing categories in hand-built indexes, such as Yahoo! or Lycos. Third, users can explore the Web by browsing (following hyperlinks that seem interesting to them). Finally, users can provide a profile of their interests to an "information push" system and hope relevant new information will be forwarded to them when it becomes available. Despite this range of possibilities, information access remains extremely tiresome and difficult, and offers users only limited glimpses into the breadth of relevant available information.

The Aware software allows the user to control and define how the search is performed. It learns preferences and cuts tedious, time-consuming steps out of the research process. It also assists users in searching online information sources by pursuing search paths that the user did not necessarily recognize or is unable to pursue due to time or skill limitations. So, while it weeds out extraneous documents that the user would eliminate on his own—a time-consuming, frustrating process akin to navigating rush hour on the "information superhighway"—it also manages to spot sources that the user might pass over, like an excellent roadside diner that the "driver" might otherwise speed by.

The Aware software loads onto any Microsoft Windows personal computer, and the program can be run from the desktop icon or from the Start menu. It is simple to get started, as the program is set up like a typical search engine, with a text field for the searcher to enter keywords. Similarly, search results display as they would with a conventional search engine.

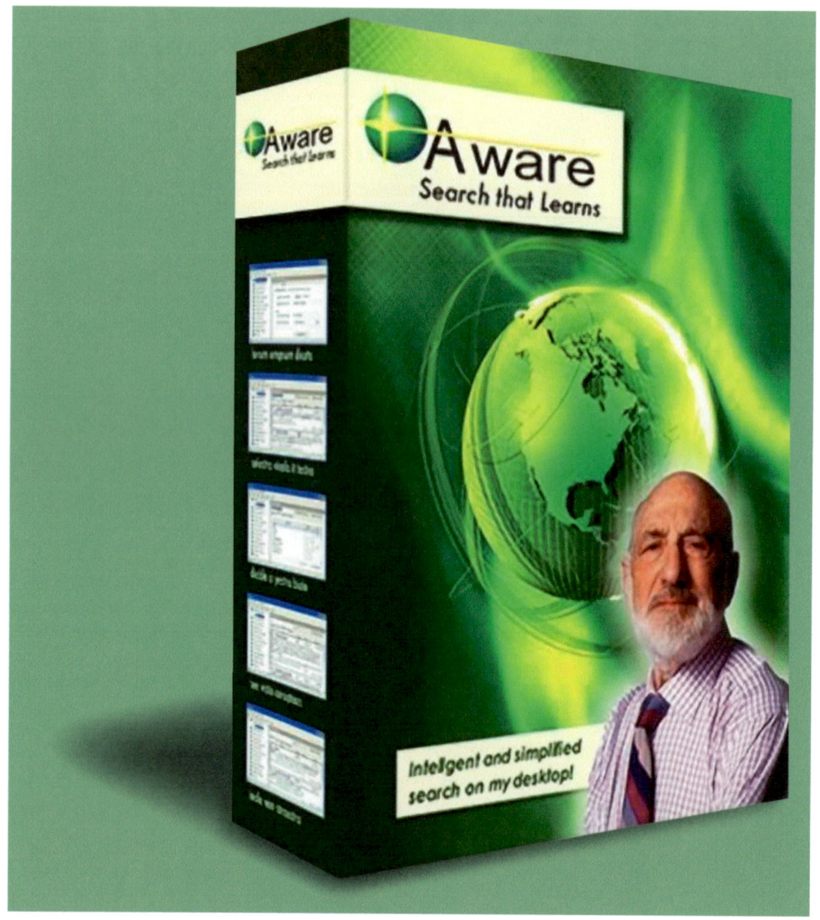

Aware, originally a NASA project designed to help its scientists perform better searches of the Internet, is making the "information superhighway" more navigable for researchers everywhere.

The software loads onto any Microsoft Windows personal computer and begins "learning" after its first search.

The user, however, can rate the types of results with a plus or minus button, a feature not found in other search engines. The software stores this information and uses it in future searches. The user also has the option of choosing to ignore a specific result or type of result. This cuts down on unwanted results being duplicated.

After checking the results of a search and categorizing them according to preference, the user can continue searching—only this time with an added advantage: the search engine remembers the preferences and tailors the search. The next set of results is targeted to the user's topic and filtered to avoid duplicates. The software will save these search parameters, so that subsequent searches will benefit from the previous queries.

Research can get more specific and focused over the course of time, all the while streamlining and controlling redundancies. The search and its parameters can be saved and the user can come back time and again. The user also has the option of creating different searches, with different levels of restrictions and filters, and with different keywords. One user can create several different types of targeted search tools.

Aware saves all searched keywords, catalogs the results collections, and learns which type of documents best suits the research goals; then it expands and filters new searches based on past search preferences. The software does all of this for the user with a minimum of input.

If the user does want to become more involved, though, he can access a terms list—created by the software, based on previous searches—to adjust the weight given to each term. With a sliding scale, the user can drag a pointer to indicate the importance one search term should be given over another. This is not the only way to refine the search. Users of Aware can change and modify search properties by choosing which sources to either highlight or ignore, by selecting the number of queries, and by setting the number of results displayed. Users can even select how many results Aware should analyze on every search.

This software is not restricted, however, to excluding search items. If the user finds something that does not fit the current search but that looks interesting, he can save search results that are off topic in a different file to pursue later. The software has yet another added feature: the ability to let multiple users access a specific search.

Aware, originally a NASA project designed to help its scientists perform better searches of the Internet, is making the "information superhighway" more navigable for researchers everywhere. ❖

Aware™ is a trademark of Stottler Henke, Inc.
Yahoo!® is a registered trademark of Yahoo! Inc.
Lycos® is a registered trademark of Carnegie Mellon University.
Microsoft® and Windows® are registered trademarks of Microsoft Corporation.

Fault-Detection Tool Has Companies 'Mining' Own Business

Originating Technology/ NASA Contribution

A successful launching of NASA's Space Shuttle hinges heavily on the three Space Shuttle Main Engines (SSME) that power the orbiter. These critical components must be monitored in real time, with sensors, and compared against expected behaviors that could scrub a launch or, even worse, cause in-flight hazards.

Since 1981, SSME faults have caused 23 scrubbed launches and 29 percent of total Space Shuttle downtime, according to a compilation of analysis reports. The most serious cases typically occur in the last few seconds before ignition; a launch scrub that late in the countdown usually means a period of investigation of a month or more. For example, during the launch attempt of STS-41D in 1984, an anomaly was detected in the number three engine, causing the mission to be scrubbed at T-4 seconds. This not only affected STS-41D, but forced the cancellation of another mission and caused a 2-month flight delay.

In 2002, NASA's Kennedy Space Center, the Florida Institute of Technology, and Interface & Control Systems, Inc., worked together to attack this problem by creating a system that could automate the detection of mechanical failures in the SSMEs' fuel control valves.

Partnership

Indialantic, Florida-based Interface & Control Systems was awarded a Kennedy **Small Business Technology Transfer (STTR)** contract to develop the failure-detection system. Using data-mining techniques, the company is able to extract behavioral characteristics from past performance data and automate real-time monitoring and controlling with a high degree of accuracy.

What resulted from the STTR was the combination of two cross-cutting technologies: Florida Institute of Technology developed adaptive machine learning algorithms which interact with graphical front-end tools and an artificial intelligence tool developed by Interface & Control Systems's. This tool, known as Spacecraft Command Language (SCL) software, is a rule-based expert software system used for real-time monitoring and control. Utilizing the adaptive machine learning algorithms, Interface & Control Systems can mine historical data to define relationships and discover operational signatures for groups of related sensors.

With the SCL expert technology, the company can build data-processing control systems that run on embedded systems for autonomous applications in a wide variety of environments. These systems are capable of monitoring thousands of rapidly changing sensors and directing hundreds of actuators based on complex, machine-generated logic.

Prior to the STTR partnership, there was no evidence of cooperation between a rule-based expert system and adaptive machine learning technology within the production, launch, or flight environments of the aerospace industry, according to Interface & Control Systems. This new offering vastly improves NASA's abilities to autonomously conduct SSME analysis in real time, while still keeping humans "in the loop"—minus the hassle of having to carry out mundane, manual tasks that can easily be handled by the failure-detection technology.

A remote camera captures a close-up view of a Space Shuttle Main Engine during a test firing at the Stennis Space Center. In addition to pre-flight testing, NASA is capable of monitoring the performance of Space Shuttle Main Engines during flight, allowing for real-time, autonomous detection of potential mechanical failures.

Product Outcome

Built on top of the SCL architecture, Interface & Control Systems's SensorMiner is a time-series data-mining tool that uses past performance data to build human-readable models for real-time fault detection. SensorMiner "learns" from the past performance data, develops a "signature" for nominal operation, and automatically generates a fault-monitoring system. Conventional technology has made this process both a labor-intensive and error-prone task that is often misunderstood, even by experts, according to Interface & Control Systems. SensorMiner, on the contrary, will automatically correlate seemingly unrelated data and provide rich graphical feedback to the user.

As part of a complete toolset, SensorMiner uses a combination of time warp, cluster, rule induction, Euclidean error, and state machine technologies. The result is a sophisticated temporal machine learner for real-time anomaly detection. Using time-series anomaly detection, the toolset trains a model on known good data, estimates the probability distribution, and assigns a likelihood-based score to new sensor data. The score is used to determine if real-time data have deviated from the signature for these data points. Errors, or "out of family" characteristics, are detected by the SCL expert system and corrective measures can be automatically invoked.

"The ability to automatically make predictions or help people make decisions faster and more accurately, in real time, has far-reaching implications that spread across industry boundaries," said Brian Buckley, vice president of marketing for Interface & Control Systems. "We designed SensorMiner to identify anomalies in a wide variety of systems."

SensorMiner's toolset also consists of data-mining tools and a real-time monitoring system. The data-mining tools interact with a graphical front end that allows a user to display data signatures discovered by the data-mining strategies. Once a signature is captured, additional data sets can be played through a built-in simulator to exercise the monitoring system and reveal graphical results. Using SensorMiner, data-mining algorithms can ultimately be fine-tuned to increase the likelihood of detecting anomalies in the real-time system.

Interface & Control Systems software engineer Walter Schiefele has contributed nearly 2 years of research and development to the SensorMiner product. "Many artificial intelligence programs yield models that are not understandable by humans," according to Schiefele. "Our approach was to generate models that are easily readable and graphically displayed. People can visualize the rules. More importantly, domain experts can manually trim or adjust rules in the final product."

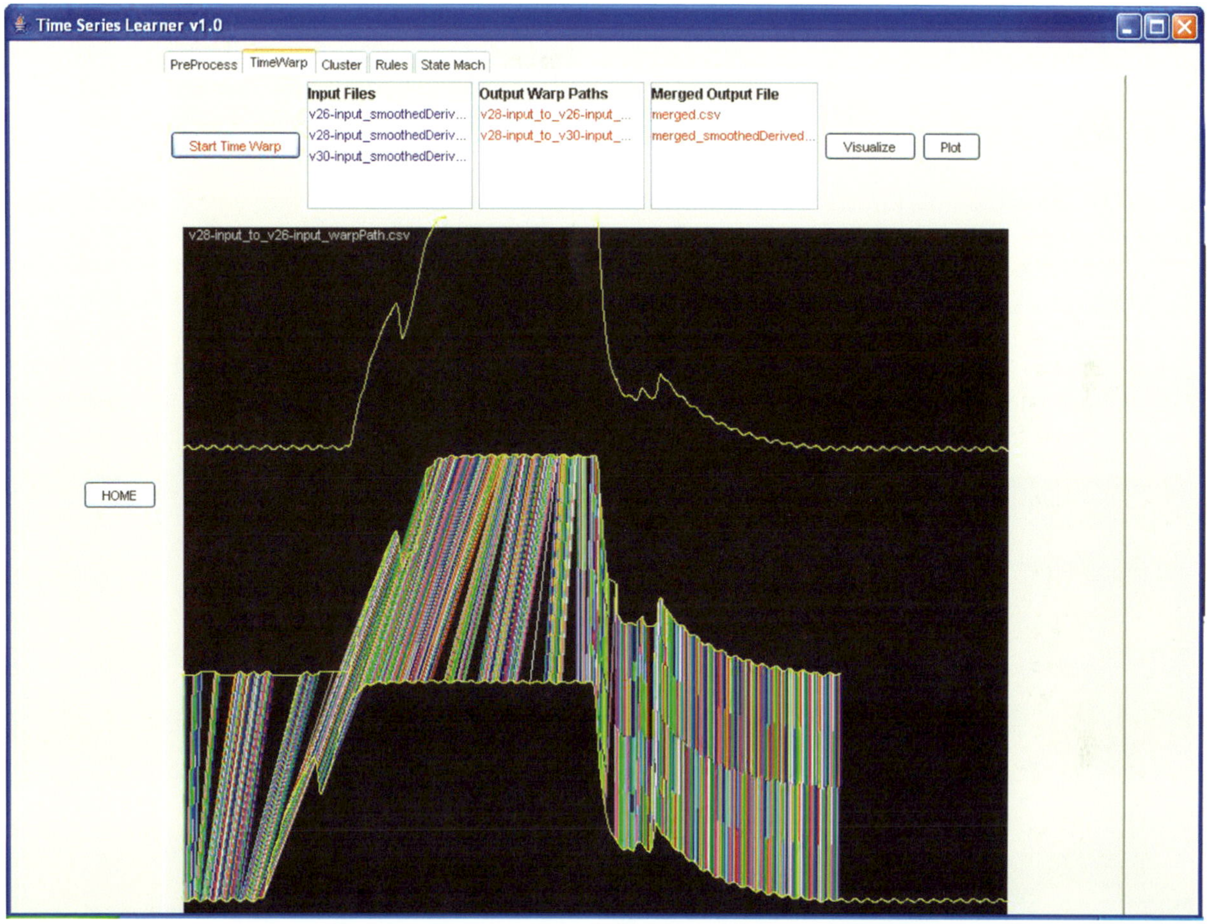

Interface & Control Systems, Inc.'s SensorMiner is a time-series data-mining tool that uses past performance data to build human-readable models for real time fault detection.

The company is introducing the commercially available data-mining tool to existing control system and test set markets that require advanced, real-time anomaly detection and analysis systems. SensorMiner will then be introduced to supervisory control and data acquisition and process-control system fields. The company is also inviting inquiries for partnerships and value-added resellers, for targeted commercial markets. ❖

Software to Manage the Unmanageable

Originating Technology/ NASA Contribution

In 1995, NASA's Jet Propulsion Laboratory (JPL) contracted Redmond, Washington-based Lucidoc Corporation, to design a technology infrastructure to automate the intersection between policy management and operations management with advanced software that automates document workflow, document status, and uniformity of document layout. JPL had very specific parameters for the software. It expected to store and catalog over 8,000 technical and procedural documents integrated with hundreds of processes. The project ended in 2000, but NASA still uses the resulting highly secure document management system, and Lucidoc has managed to help other organizations, large and small, with integrating document flow and operations management to ensure a compliance-ready culture.

Partnership

In the mid-1990s, NASA was pushing toward process-based-management for documents and procedures. John Casani, JPL's chief engineer, was given the assignment of implementing this new system, but after scouring the market for commercially available software, he found none that was powerful, comprehensive, or secure enough for the task of providing more than 8,000 documents to over 7,000 users located on seven continents, and that would allow JPL to become ISO 9000 certified.

JPL partnered with Lucidoc to create the software solution. NASA employed a service contract to streamline the process, and the result was an efficient partnership and a quick, thorough solution to the problem.

JPL uses the Lucidoc solution to create a living database of ever-changing and highly integrated procedural information. The software is an HTML, object-oriented, workflow management system that controls guidelines and standards across the laboratory and the various campuses. One of the key advantages is that the user can access the system through a standard Internet browser.

Product Outcome

Having proven successful at cataloging, organizing, and maintaining the wealth of information NASA required, Lucidoc was uniquely prepared to offer the same services to a variety of other industries. Lucidoc has been effective in the banking and health care industries, as well as in the government aerospace industry.

The Lucidoc system automates the operational life cycle of document management and compliance with several key features. It offers multilevel communication of information between departments and staff, and fosters collaboration between departments and facilities on feedback, changes in documents or procedures, and streamlining the approval and sign-off stages of document flow. It provides document creation and revision capabilities, including links to related documents requiring changes. Unlike typical document management systems, which simply delete outdated documents, Lucidoc archives and updates files, making it a living document system with an active history.

The Lucidoc system enables an organization to affirm that compliance information is distributed, authoritative, consistent, complete, and managed.

The information in the Lucidoc system is automatically distributed to all of the appropriate people, which

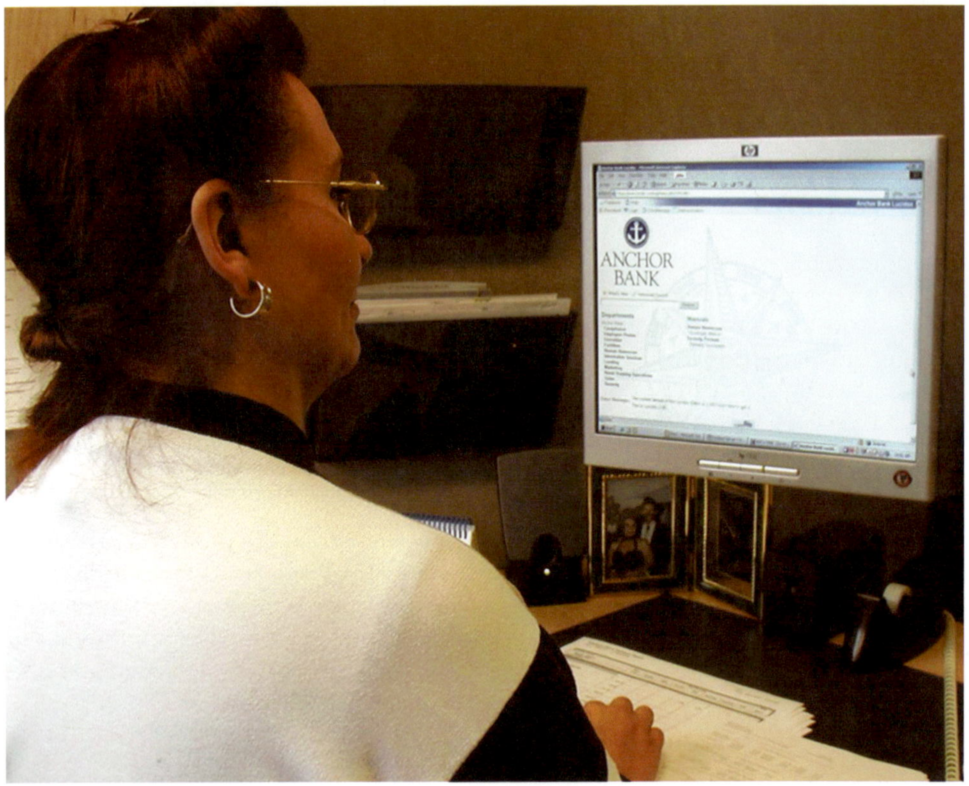

Employees at Anchor Bank, a state-chartered mutual savings bank, headquartered in Aberdeen, Washington, are divided over 14 different branches and across 18 separate facilities, yet they are united by the Lucidoc document management system.

is one of the key goals of any document management system, because without access to the information, the process stalls, requiring time-consuming delays. In many instances, the Lucidoc system replaces binders and files that are difficult to keep current and oftentimes accessible only to a few people in specific locations.

Furthermore, the information is authoritative. It is constantly updated and made relevant, and it is the same information available to every other user on the system. It has been checked and validated; the right people have been given the opportunity to sign off when needed, thereby giving the necessary approval.

The information found on the system is consistent, because it is all regulated within the same document management system. The possibility of redundancy is reduced, and the likelihood of inconsistent answers or directions is eliminated.

In addition to being distributed, authoritative, and consistent, the information in the system is complete. It links like information to like information, and because of its Web-based approach, the system allows for thorough searching.

Finally, it is life-cycle managed. Since this is an active document management system, it accommodates revisions and changes in documents, as well as adaptations in processes.

The many facets of this document management system make it ideal for use by organizations seeking to make their procedures and documents uniform across several offices. Just as it has found a long-term home at JPL, this specialized software has found applications in the document-heavy industries of banking and medicine, and it has proven effective and helpful in both.

Community banks and credit unions are regulated by their state governments, the Federal Government, and groups such as the U.S. Securities and Exchange Commission, Fair Housing Regulators, and insurers. These financial institutions risk hefty civil and criminal penalties for compliance failures. Banks must be able to show their regulators conformity and compliance through regular performance audits and prospective evaluations of their documentation. As banks move toward more and more micro-branches, like ones seen at grocery stores, this becomes increasingly difficult. The Lucidoc system provides a solution for keeping documents up to date, distributed, and in compliance with constantly changing regulations.

The Lucidoc system is working to help medical professionals in California get the information they need when they need it.

Anchor Bank, a state-chartered mutual savings bank in Washington State, employs the Lucidoc system to provide document management services to its 18 facilities and 14 branches. With so many locations, sending out documents to each of the offices was time-consuming and costly. The bank wanted a secure system that would automate the creation of and access to operational documents. After researching the alternatives, Anchor Bank realized that Lucidoc was uniquely suited to solve its problems. The system streamlined the document workflow process, reduced the cost of relying on paper documents, and sped the spread of knowledge.

Two additional advantages came out of this partnership: Anchor Bank now has virtually stress-free audits, as all of its paperwork is automatically updated; and, because of its intuitive, easy-to-use format, employees are finding the Lucidoc portal useful for human resources materials and onsite training.

Lucidoc is making similar inroads in the hospital industry. The Lucidoc system provides health care staff with instant access, from any location, to the accurate procedural information and care directions they need. Information access and accuracy are the keys to providing consistent, quality care.

The Joint Commission on Accreditation of Healthcare Organizations requires that policies and protocols be kept up to date, and the Lucidoc solution allows hospital leadership to automate the review and updating processes so that staff can concentrate on quality management and care, rather than on performing an endless paper shuffle.

Lucidoc is currently being used by Palomar Pomerado Health, the largest public health district in California, with over 800 miles of territory. This district includes the 326-bed Palomar Medical Center, the 115-bed Pomerado Hospital, 2 skilled nursing facilities, 5 health centers, a surgery center, a home health program, physician referral services, and wellness classes. Prior to Lucidoc, all of the documents needed to run these facilities were managed by hand. In fact, it was one person's job to coordinate all documents for review.

The Lucidoc system provides efficient transfer of information, including electronic signatures, a near must in the medical field. It also provides nursing teams with access to an up-to-date, authoritative database with intuitive search capabilities. For example, if a nurse needs information about an obscure procedure, rather than hunt down a manual, which may be out of date or even in a locked cabinet, she can type several key words into a browser, as if she were performing a Google search. Unlike with a traditional Internet search, though, the nurse can be assured that the information she accesses is current and authoritative, because it is within the Lucidoc system. ❖

Lucidoc™ is a trademark of the Lucidoc Corporation.
Google™ is a trademark of Google Inc.

Tracking Electromagnetic Energy With SQUIDs

Originating Technology/ NASA Contribution

A superconducting quantum interference device (SQUID) is a gadget used to measure extremely weak signals, specifically magnetic flux. It can detect subtle changes in energy, up to 100 billion times weaker than the electromagnetic energy required to move a compass needle. SQUIDs are used for a variety of testing procedures where extreme sensitivity is required and where the test instrument need not come into direct contact with the test subject.

NASA uses SQUIDs for remote, noncontact sensing in a variety of venues, including monitoring the Earth's magnetic field and tracking brain activity of pilots. Scientists at NASA's Goddard Space Flight Center have been making extensive use of this technology, from astrophysical research, to tracking the navigational paths of bees in flight to determine if they are using internal compasses.

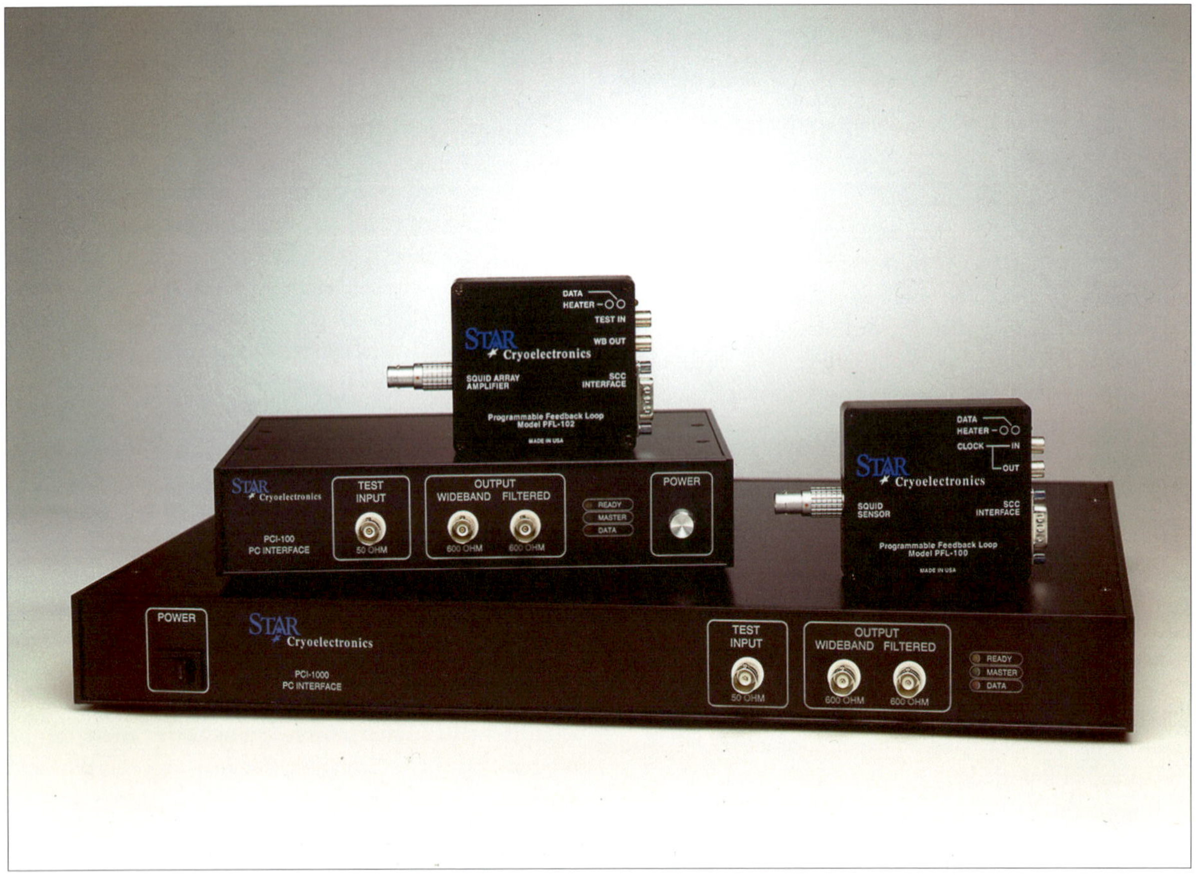

STAR Cryoelectronics, LLC's line of single- and multi-channel PC-based readout electronics for SQUIDs and SQUID amplifiers.

These very sensitive measurement devices have a wide variety of uses within NASA and even more uses within the commercial realm.

Partnership

STAR Cryoelectronics, LLC (STARCryo), of Los Alamos, New Mexico, specializes in developing, manufacturing, and marketing ultrasensitive SQUID sensors and advanced personal computer-based SQUID control electronics. Its customer base is worldwide.

STARCryo was successful in obtaining NASA Phase I and II **Small Business Innovation Research (SBIR)** contract funding to develop new technologies and bring them to market. This funding included performing advanced electronics and sensor work at Goddard. The work involved generating voltage gain through arrays and cryogenic amplifiers for detectors.

Product Outcome

STARCryo's SQUID is the most sensitive detector of magnetic flux available, making it ideal for a variety of uses, including high-resolution measurements of current, voltage, magnetic field or field gradient, gravitational field, and magnetic susceptibility. STARCryo now offers an extensive range of SQUID sensors and packaging options for applications in biomedical imaging, nondestructive testing of materials, geophysical exploration, and basic research.

STARCryo's SQUIDs are in use at major university, government, and corporate research laboratories all throughout the world.

The PC-based SQUID control-electronics that STARCryo manufactures are marketed under the trade name pcSQUID and offer unsurpassed flexibility and performance for single- and multi-channel applications, in addition to being convenient and easy to use. A convenient interface for Microsoft Windows puts the SQUID control on the computer, fully integrated with data acquisition and analysis tools.

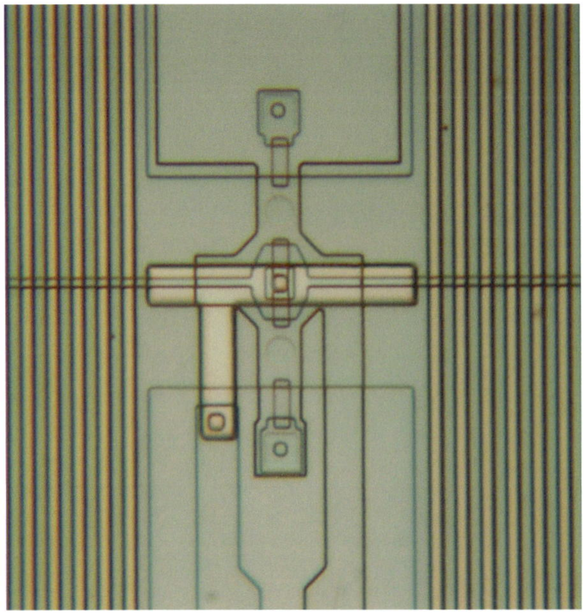

The input stage of a SQUID amplifier.

STARCryo's sensors are used at medical research centers around the world to record and localize the magnetic signals associated with neural activity in the brain. Magnetoencephalography (MEG), the monitoring of magnetic fields in the brain, is emerging as the primary tool for investigating brain wave activity. Other competing techniques, such as magnetic resonance imaging (MRI) and positron emission tomography (PET), cannot offer the millisecond temporal resolution attainable with MEG.

MEG is an essential tool for mapping functional brain activity and may be superposed with MRI data in order to correlate localized brain activity with anatomical information. The data are invaluable for pre-operative surgical planning, and insurance companies in the United States are now paying reimbursements for the procedure.

STARCryo's pcSQUID is also being used with magnetocardiography (MCG), a modern medial approach to monitoring heart health, primarily a patient's risk of sudden cardiac death, a disease that claims over 300,000 lives in the United States each year. STARCryo's ultra-sensitive SQUID sensors are ideally suited for applications such as MCG.

The SQUID sensors have also found applications in the field of nondestructive evaluation (NDE), where they are ideal, due to their extraordinary sensitivity and enormous bandwidth. The NDE applications fall into three categories: imaging of surfaces and sub-surface features, flaws, defects, and voids in materials; imaging of materials that have been pre-magnetized by a pulsed or static background magnetic field; and imaging of intrinsic or applied electric currents in electronic circuits or materials.

In the aerospace industry, SQUIDs are being used to obtain magnetic images of cracks in aircraft wheel rims and to inspect jet engine turbine blades, as well as to image defects in models of typical aircraft lap joints. Other applications include the evaluation of steel plates used to build ship hulls and the inspection of concrete infrastructures, such as bridgework.

STARCryo has also found use for their SQUIDs in the field of geophysical exploration. Again, due to their extreme sensitivity, small size, enormous bandwidth, and flat frequency response, they have been used for bore hole depth profiling, airborne surveying, transient electromagnetics, magnetoellurics, and controlled source audio magnetoellurics. ❖

pcSQUID™ is a trademark of STAR Cryoelectronics, LLC.
Microsoft® and Windows® are registered trademarks of Microsoft Corporation.

A process engineer depositing thin films used to fabricate SQUID devices.

Taking the Risk Out of Risk Assessment

Originating Technology/ NASA Contribution

The ability to understand risks and have the right strategies in place when risky events occur is essential in the workplace. More and more organizations are being confronted with concerns over how to measure their risks or what kind of risks they can take when certain events transpire that could have a negative impact.

NASA is one organization that faces these challenges on a daily basis, as effective risk management is critical to the success of its missions—especially the Space Shuttle missions.

On July 29, 1996, former NASA Administrator Daniel Goldin charged NASA's Office of Safety and Mission Assurance with developing a probabilistic risk assessment (PRA) tool to support decisions on the funding of Space Shuttle upgrades. When issuing the directive, Goldin said, "Since I came to NASA [in 1992], we've spent billions of dollars on Shuttle upgrades without knowing how much they improve safety. I want a tool to help base upgrade decisions on risk." Work on the PRA tool began immediately.

The resulting prototype, the Quantitative Risk Assessment System (QRAS) Version 1.0, was jointly developed by NASA's Marshall Space Flight Center, its Office of Safety and Mission Assurance, and researchers at the University of Maryland. QRAS software automatically expands the reliability logic models of systems to evaluate the probability of highly detrimental outcomes occurring in complex systems that are subject to potential accident scenarios.

Even in its earliest forms, QRAS was used to begin PRA modeling of the Space Shuttle. In parallel, the development of QRAS continued, with the goal of making it a world-class tool, one that was especially suited to NASA's unique needs. From the beginning, an important conceptual goal in the development of QRAS was for it to help bridge the gap between the professional risk analyst and the design engineer. In the past, only the professional risk analyst could perform, modify, use, and perhaps even adequately understand PRA. NASA wanted to change this by developing a PRA tool that would be friendlier, more understandable, and more useful to the broader engineering community. This concept ultimately led to the look, feel, and functionality that QRAS has today.

The Quantitative Risk Assessment System (QRAS) is a comprehensive tool for conducting probabilistic risk assessment.

Partnership

In July 2003, Item Software (USA) Inc., of Anaheim, California, received an exclusive license for the QRAS software. The company is a leader in providing software solutions and services for reliability, availability, maintainability, safety, quality assurance, and risk assessment to government and commercial customers in aerospace and other industries. It built on the already-state-of-the-art features of QRAS, preparing the software for commercial sale as part of its suite of software and associated services. As part of the commercialization and distribution of QRAS, Item Software also supplies support and training.

The license provides the basis for a potential partnership between NASA and Item Software for the future extension and deployment of QRAS and related software, as may be needed for NASA purposes.

Product Outcome

Item Software developed the first commercialized version of QRAS and released it in January 2005. The new, enhanced QRAS 1.80 provides a cohesive and simple-yet-powerful platform for system risk assessment, through a large range of failure probability characterizations for engineering and scenario-driven applications.

The software builds a risk model of a system or a scenario for which risk of failure is being assessed, then analyzes the associated risk to the risk model. It can then be used to perform sensitivity analysis of the risk model by altering fundamental components and quantification models. During this stage, a fixed baseline is constructed and stored. This baseline contains the solutions for the lowest level scenarios, preserved in an event tree structure. The analysis, at any level of the hierarchy, aggregates these baseline results for risk quantitative computation as well as ranking of a particular risk.

The commercial version introduces substantial enhancements and includes many new features not seen in the original version. For example, it contains fault tree analysis, a deductive procedure for determining the various combinations of hardware and software failures, plus human errors that could result in the occurrence of specified undesired events. Fault tree analysis is one of the most widely used methods in system reliability analysis.

A new fault tree editor with extensive editing and user-friendly features was also added to the commercial version to allow users to effortlessly create, review, and analyze multiple fault trees simultaneously. The fault trees created with the fault tree editor can be attached to the event sequence diagram (ESD) in QRAS projects. An ESD is a visual representation of a set of possible outcome scenarios originating from an event. Each scenario in an ESD consists of a unique sequence of pivotal events, and eventually leads to an end-state that designates the severity of the outcome of a particular scenario.

QRAS assesses risk at the failure mode, subsystem, and element (i.e., a group of subsystems) levels, based upon user-supplied quantification of failure models, event sequence system decompositions, and system operating time. It provides users with structured guidance so managers, engineers, and even individuals who may not be experienced in the field of risk assessment can use it.

Overall, the QRAS software's unique, patented PRA capabilities assist risk analysts in modeling deviations from a system's nominal functions, the timing and likelihood of such deviations, potential consequences, and scenarios leading from initial deviations to such consequences.

The U.S. Department of Homeland Security, The Boeing Company, Booz Allen Hamilton, Inc., ITT Industries, Inc., the European Space Agency, and the China National Space Administration are just some of the entities evaluating QRAS for their risk assessment needs.

Those currently employing it—other than NASA—include the Canadian Space Agency, AES Corporation, and Harvard Medical School. At Harvard Medical School's major teaching hospital, Beth Israel Deaconess Medical School Center, QRAS is being utilized to determine the health care risk associated with general surgical processes and surgical devise use, pharmaceutical ordering, transfusion services, and organ procurement and transplantation.

Other examples of application include determining the probability of airplane crashes arising from factors such as engine failure, avionics failure, or human-failure at the air control tower; and train collisions caused by failures in train-signaling systems. ❖

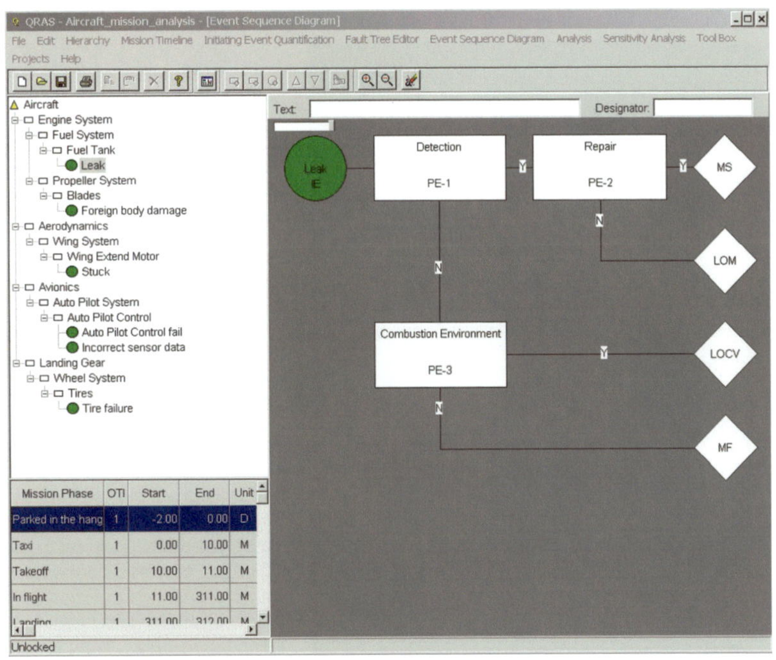

QRAS provides a full graphical user interface, including fault tree editors and event sequence diagrams, plus analysis screens that provide results in tabular and graphical formats. The screen to the left portrays System Hierarchy Mission Phase and event sequence diagrams that are used to determine the consequence of an initiating event and the expected frequency of each consequence. For example, a leak in an aircraft fuel system could result in the following user-defined end states: Mission Success (MS) if the leak is repaired, or Loss of Mission (LOM), Loss of Crew and Vehicle (LOCV), or Mission Failure (MF).

Satellite and Ground System Solutions at Your Fingertips

Originating Technology/ NASA Contribution

In the summer of 1998, the blockbuster action movie "Armageddon" captivated audiences with a thrilling doomsday plot about a meteor the size of Texas that was racing towards the Earth. Though the premise of the movie was purely fictional, the unfortunate reality is that near-Earth asteroids such as the one portrayed in the film do exist.

On December 23, 2004, NASA announced that an asteroid it anticipated to pass near the Earth on April 13, 2029, had been assigned the highest score to date on the universally used Torino Impact Hazard Scale. At first, the flyby distance for the asteroid, dubbed MN4, was uncertain and an Earth impact could not be ruled out. The odds of impact were initially believed to be 1 in 300, high enough to merit special monitoring by astronomers around the world, but were then escalated to 1 in 37 on December 27. NASA officials noted, however, that these odds should not be of public concern, since they were likely to change on a day-to-day basis as new data were received. The officials were correct in their assertion, as any chances of an impact with Earth in 2029 were completely ruled out later that same day.

Integral Systems, Inc., a leading provider of satellite ground systems and the first company to offer an integrated suite of commercial-off-the-shelf software products for satellite command and control, is helping NASA keep a careful watch for any close-encountering asteroids with its tracking technology. The company supported the first NASA Discovery mission, the Near Earth Asteroid Rendezvous (NEAR) program, back in 1996, and has expanded its business by building more ground systems for a greater variety of satellites than any other company in the world. (NASA has since launched seven more Discovery missions, with the eighth lifting off earlier this year.) The experience gained from the company's participation in developing satellite command and control ground systems for the NEAR program has bolstered its flagship product line, the EPOCH Integrated Product Suite (IPS), first featured in *Spinoff* 1997, and led to the creation of its latest product, the Skylight Direct Broadcast Ground Terminal.

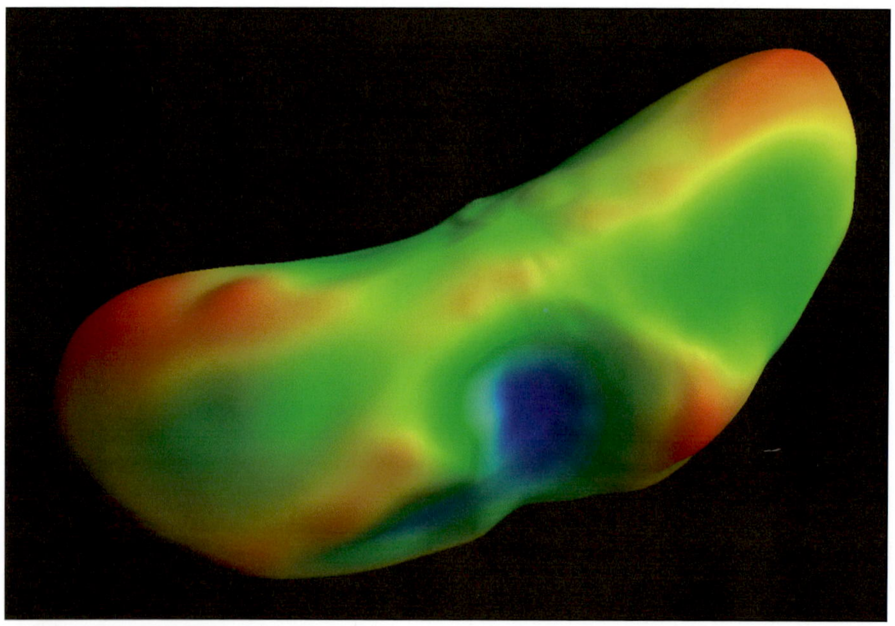

Asteroids that come within 121 million miles (195 million kilometers) of the Sun are known as near-Earth asteroids. Asteroid 433 Eros, pictured here as a topographical shape model, is one of the largest near-Earth asteroids.

Partnership

In supporting the NEAR program, Integral Systems was selected by the Johns Hopkins University Applied Physics Laboratory—the manager of this NASA Discovery mission—in November 1996 to provide up to 15 Low-Earth Orbit Autonomous Ground Terminals (LEO-Ts) for NASA's Goddard Space Flight Center. The LEO-Ts were designed to make it easier and less expensive for principal investigators to obtain telemetry, tracking, and control services for the NEAR mission and all other scientific endeavors managed by the Applied Physics Laboratory. Integral Systems's work on this project was directly incorporated into its commercial offering, the EPOCH IPS. Additionally, Integral Systems's continued work with Goddard has entitled it to adopt some of NASA's latest direct readout scientific algorithms for its new Skylight product.

Product Outcome

Satellite operators spend billions of dollars ensuring that orbiting satellites do not come crashing down, and millions more just to send simple commands to them. Integral Systems's EPOCH IPS presents a more efficient, economical alternative to costly command and monitoring practices, as an all-in-one orbital command and control system that can be operated right from a user's desktop.

Now that's managing space from the workspace.

According to the company, every single U.S. weather satellite operator and more than half of all global commercial satellite operators depend on EPOCH IPS to manage their satellites.

"We've become the *de facto* industry standard for satellite operations," said Steve Chamberlain, Integral Systems's chief executive officer. "We've got the only software products that come right out of the box and fly on any satellite from any manufacturer."

Customers employing EPOCH IPS can go from controlling one spacecraft to another without ever leaving their desk to go to another computer. The company compares the comprehensive EPOCH technology to a large, universal remote control; by simply pointing and clicking at any controllable commercial satellite whose information is stored in a computer system, EPOCH IPS can fly it right from the desk chair in front of that computer.

Government clients include NASA, the U.S. Air Force, and the National Oceanic and Atmospheric Administration, while commercial clients include Loral Skynet, PanAmSat Corporation, and New Skies Satellites.

Following on the success of the EPOCH suite, Integral Systems recently introduced the Skylight Direct Broadcast Ground Terminal. Skylight was developed to provide a complete architecture for the reception and processing of remote-sensing data collected from low-Earth orbiting satellites. Some of these data are processed to create fire-detection maps, volcano-eruption alerts, and atmospheric profiles for sea surface temperature and vegetation indexes.

The Skylight terrestrial-based terminal provides full end-to-end image-processing capabilities, including data acquisition and processing. All of the capabilities are offered within a framework that allows the system to evolve easily and cheaply. For example, new missions can be easily added, and the system can be reconfigured to support a larger image-processing enterprise. Additionally, geographic information system-based visualization tools have been incorporated to allow searches of archived images and their associated metadata, and a Web-based data distribution process greatly enhances the system's ability to share data among user groups anywhere in the world.

Integral Systems, Inc.'s satellite ground system facility. According to the company, it has built more ground systems for a greater variety of satellites than any other company on Earth.

In late 2003, the Institute for the Application of Geospatial Technology (IAGT) at Cayuga Community College, Inc., in Auburn, New York, installed a complete Skylight ground system for receiving and processing Earth science data. The system was purchased through a partnership of Integral Systems with Global Science and Technology, Inc., of Greenbelt, Maryland. It is currently receiving, processing, and storing data from the Moderate Resolution Imaging Spectroradiometer (MODIS) instrument aboard NASA's Terra and Aqua satellites, two of the preeminent aerial transmitters belonging to NASA's Earth Observing System. IAGT has a wide-ranging mission to develop educational, scientific, and economic applications of geospatial technologies, particularly in the northeastern United States and in the Finger Lakes region of upstate New York, where it is situated. Access to MODIS data affords the institution great opportunities, such as broadening aircraft-based soil and vegetation analyses that bear on the Finger Lakes grape harvest, a major economic contributor in this winemaking region.

Skylight is now being repositioned to address military situation-awareness applications. According to the manufacturer, Skylight terminals could provide field commanders and war fighters with up-to-date weather information, taking the guesswork out of deploying troops, military resources, and laser-guided munitions in fast-changing weather scenarios.

Both the EPOCH IPS and Skylight product lines offer end-to-end capability centered on highly distributed, open-architecture design concepts. These features allow complete systems to be implemented rapidly and at a very low cost, backing Integral Systems's motto: "Out-of-this-world solutions with down-to-Earth pricing and automation, delivering an entirely new world of data." ❖

Structural Analysis Made 'NESSUSary'

Originating Technology/ NASA Contribution

Everywhere you look, chances are something that was designed and tested by a computer will be in plain view. Computers are now utilized to design and test just about everything imaginable, from automobiles and airplanes to bridges and boats, and elevators and escalators to streets and skyscrapers.

Computer-design engineering first emerged in the 1970s, in the automobile and aerospace industries. Since computers were in their infancy, however, architects and engineers during the time were limited to producing only designs similar to hand-drafted drawings. (At the end of 1970s, a typical computer-aided design system was a 16-bit minicomputer with a price tag of $125,000.)

Eventually, computers became more affordable and related software became more sophisticated, offering designers the "bells and whistles" to go beyond the limits of basic drafting and rendering, and venture into more skillful applications. One of the major advancements was the ability to test the objects being designed for the probability of failure.

This advancement was especially important for the aerospace industry, where complicated and expensive structures are designed. The ability to perform reliability and risk assessment without using extensive hardware testing is critical to design and certification.

In 1984, NASA initiated the Probabilistic Structural Analysis Methods (PSAM) project at Glenn Research Center to develop analysis methods and computer programs for the probabilistic structural analysis of select engine components for current Space Shuttle and future space propulsion systems. NASA envisioned that these methods and computational tools would play a critical role in establishing increased system performance and durability, and assist in structural system qualification and certification. Not only was the PSAM project beneficial to aerospace, it paved the way for a commercial risk-probability tool that is evaluating risks in diverse, down-to-Earth applications.

Partnership

Southwest Research Institute (SwRI), headquartered in San Antonio, Texas, is a multidisciplinary, independent, nonprofit, applied engineering and physical sciences research and development organization. As the prime contractor on the PSAM project, the company developed a sophisticated computer program called NESSUS (numerical evaluation of stochastic structures under stress). Designed specifically for predicting structural response caused by uncertain basic variables such as loads, material properties, geometry, and boundary conditions, NESSUS is used by NASA to assist in the evaluation of existing critical Space Shuttle components, including the Space Shuttle Main Engines (SSMEs).

For example, the software examined a SSME fuel turbopump blade that was subjected to harsh conditions, which could ultimately affect its performance abilities. A finite element model of the turbopump blade was analyzed by NESSUS for the effects of high-temperature and high-cycle mechanical fatigue that could cause it to crack. The software computed a reliability value of 0.99978 for the SSME component. Furthermore, the probabilistic sensitivity factors revealed the variables for which tighter control would result in a more reliable blade design. (In this case, hot gas seal leakage clearly dominated other considerations.) Conversely, the sensitivity factors also exposed which variables are relatively unimportant in determining blade reliability—information important in establishing design and manufacturing controls for maximum cost-effectiveness as well as structural reliability.

The 10-year PSAM project officially ended in February 1995. Version 6.2 of NESSUS—regarded the final version for NASA's purposes at that time—was delivered to NASA in September 1995, along with a final report. Though in 2002, SwRI was again contracted by NASA, through Glenn, to further enhance NESSUS for application with large-scale aero-propulsion systems. Additionally, another government contract issued through the Los Alamos National Laboratory led to the utilization of NESSUS for extremely large and complex weapon-reliability problems, in support of its Stockpile Stewardship program.

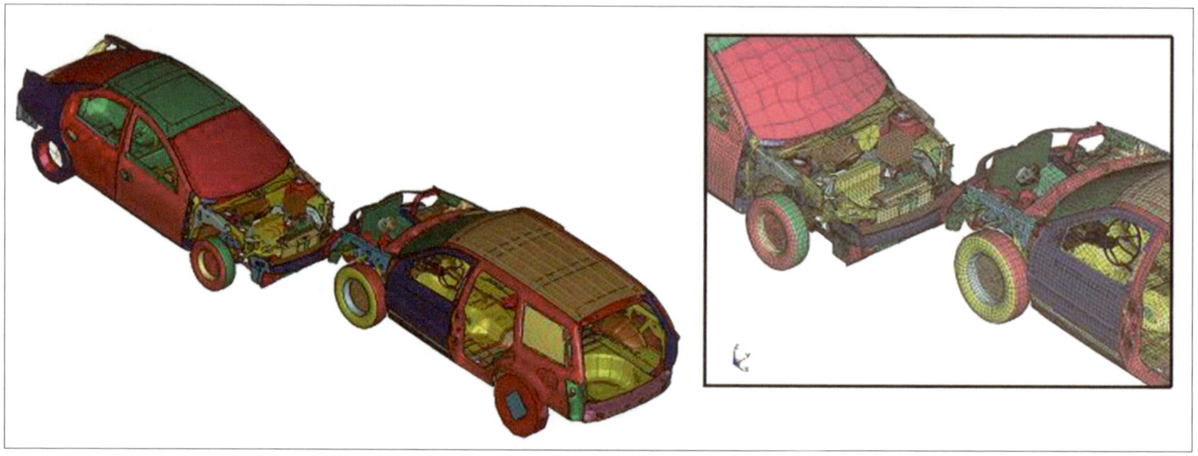

Southwest Research Institute's NESSUS software is particularly useful for large, complex calculations, such as simulating a small car colliding with a large sport utility vehicle.

The conclusion of the NASA projects, along with the Los Alamos project, marked a new beginning for SwRI, as it went on to commercialize NESSUS program for non-aerospace applications.

Product Outcome

NESSUS is a modular computer software system that integrates advanced reliability methods with finite element and boundary element methods and probabilistic algorithms in order to model uncertainties in loads, material properties, and geometries with random variables. Probabilistic performance models implemented in NESSUS include stress, strain, displacement, vibration, fatigue, fracture, and creep. NESSUS can perform reliability analyses for multiple components and failure modes, and identify critical random variables and failure modes to support structural design, certification, and risk assessment.

Version 8.2, the latest, was released to U.S. government organizations in December 2004 and to the public in 2005. While the software continues to serve the aerospace industry, it is also solving a diverse range of problems in a variety of other industries, including health and medicine, automotives, biomechanics, nuclear waste packaging, munitions (weapon systems), and offshore pipeline construction. To accomplish this, NESSUS has been interfaced with many well-known third-party and commercial deterministic analysis programs.

Recently, NESSUS provided the U.S. Naval Biodynamics Laboratory/U.S. Naval Air Warfare Center Aircraft Division with a probabilistic methodology and computational tool for evaluating the risk of cervical spine injury.

Cervical spine injuries occur as a result of impact or from large inertial forces such as those experienced by military pilots during ejections, carrier landings, and emergency crash-landings on water, also known as "ditchings." Other examples include motor vehicle, diving, and athletic-related accidents. Reducing the likelihood of occupant injury by identifying and understanding the primary injury mechanisms and the important factors leading to injury motivates most research in this area. Therefore, it is of broad interest to SwRI to design occupant safety systems that minimize the probability of injury.

To do this, the company says, the designer must have quantified knowledge of the probability of injury, due to different impact scenarios, and also know which model parameters contribute the most to the injury probability. Although complex numerical (e.g., finite element) models are becoming more widely used as a means of augmenting and extending laboratory testing, most are deterministic in that they do not quantify the effect of uncertainties on the computed model responses.

In developing a probabilistic methodology and computational tool for performing the spinal evaluation, calculations were generated using the NESSUS probabilistic analysis program in conjunction with the commercial finite element program, ABAQUS. The company intends to use the resulting models for additional applications, including the study of spinal behavior under normal and distressed conditions, the design of implants utilizing novel materials and/or configurations, analysis of novel instrumentation systems that may help avoid costly experimentation, and the design of anthropomorphic test devices or physical models that replicate human response such that injury under simulated dynamic conditions can be replicated.

In helping NASA return to flight, NESSUS 8.2 was recently used to solve a critical Space Shuttle problem for the Agency's Engineering Safety Center, which was created in the aftermath of the Columbia accident to serve as an independent technical resource for NASA managers and employees. This project involved developing probabilistic and deterministic fracture mechanics models to quantify the reliability of a flowliner in a SSME. In another "Return to Flight" project for Johnson Space Center, the software is being utilized on a regular basis to assess the probability of foam-impact damage to the Space Shuttle's reinforced carbon-carbon leading edge, if such were to happen upon ascent.

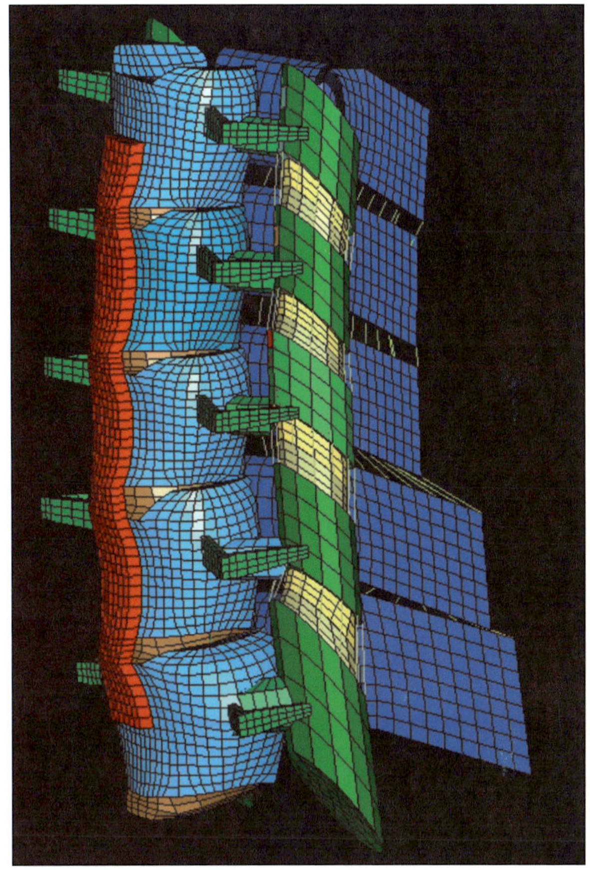

NESSUS 8.2 was used to quantify the risk of cervical spine injury to aviators during ejection.

In July, SwRI won R&D Magazine's R&D 100 Award, with NESSUS Version 8.2 being crowned as one of the 100 most significant technological achievements of the past year. The company has won 29 R&D 100 Awards since 1971. ❖

Software of Seismic Proportions Promotes Enjoyable Learning

Originating Technology/ NASA Contribution

While working for NASA, Jack Sculley and Terry Brooks had a revelation. They wanted to find a novel and unique way to present the scientific principles of NASA research to the public, so as to not only enlighten, but entertain. Suddenly, their revelation morphed into something even grander. "Why stop at NASA?" they asked themselves. With this thought, Sculley and Brooks left NASA and set out to convey voluminous scientific findings from different organizations in the form of digital, interactive media that would enhance the exploration and adventure interests of people of all ages.

Sculley, a former researcher at Ames Research Center, the Jet Propulsion Laboratory (JPL), and Apple, Inc.'s and LucasFilm Ltd.'s multimedia labs, and Brooks, a former public information officer at JPL and an Emmy award-winning documentary film producer, founded Seismic Entertainment in 1989 to communicate their "edutainment" ideas. The two acknowledge that NASA has provided much of the inspiration and content for Seismic Entertainment over the past decade and a half. Additionally, Sculley's experience as a virtual reality and Mars specialist and Brooks's experience creating NASA public access programs were significant to the San Francisco-based company's success. Its most recent project, "Inside NASA," provides virtual tours of NASA's field centers and allows for a comprehensive focus on the broad range of NASA programs for the benefit of the general public.

Partnership

In addition to the inspiration and knowledge acquired while employed by NASA, the founders gained assistance from the Agency in the form of **Small Business Innovation Research (SBIR)** contracts. They applied for a NASA SBIR grant with Marshall Space Flight Center in 1998, and were awarded a Phase I feasibility study contract and later a Phase II commercialization contract to develop virtual tours of NASA facilities, Web sites, image archives, and programs.

Together with Tom Simmons, Seismic Entertainment's producer, Sculley and Brooks visited all 10 NASA centers and several subsidiary facilities, and pored over countless NASA Web sites, data archives, and program outreach materials before selecting the final set of locations and programs to feature in the "Inside NASA" product. Photographic scouting visits to all the centers led to final location shot lists, which were developed with the centers' public information and outreach personnel.

Photography began in 2001, and a special hemispheric lens was used to create spherical panoramas of numerous locations, including rocket test chambers, launch pads, astronaut training facilities, scientific test facilities, and rover obstacle courses. Production of the virtual tours commenced that fall. The ensuing software product featured the field center tours, in addition to 10 topic tours, which followed programmatic themes across multiple centers, such as aeronautics, robotics, rocket engine development, astronaut training, and space science research.

Initial versions of each center's tour were uploaded to a private site, where NASA public information personnel could access them and provide feedback to the Seismic Entertainment staff. The final "Inside NASA" software product was released for public use on CD-ROM in late 2003, and NASA has folded the virtual tours into several new employee training and outreach Web sites.

Product Outcome

Designed for NASA, educators, and the general public, and employing the latest developments in virtual reality, "Inside NASA" is available in science center stores and retail outlets across the country. One of its many highlights is an extensive tour of Johnson Space Center, developed in collaboration with Johnson's staff. In embarking on this virtual tour, users can take a plunge

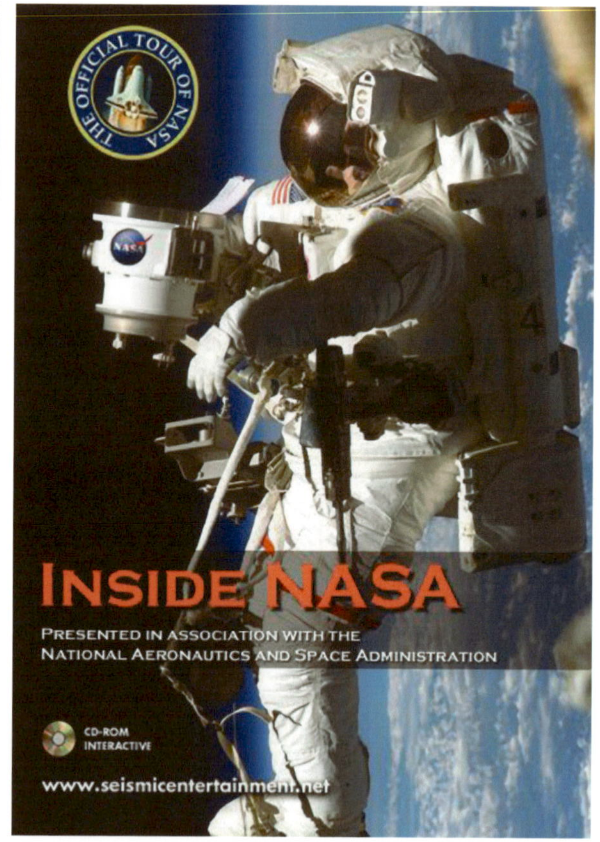

"Inside NASA" is packed with in-depth virtual tours that let users discover the inner workings of NASA, as well as the outer reaches of space.

with astronauts in the National Buoyancy Laboratory, a 22.7 million-liter swimming pool in which fully-suited astronauts spend 7 hours training for every hour they will spend spacewalking during a mission. Users can also wander through Space Shuttle and Space Station simulators, and observe experiments being performed in large, zero-gravity vacuum test chambers.

To navigate a tour, users can either move a cursor over the up, down, left, and right arrow tabs to travel in any of these directions, or click and drag within the image itself. When browsing a panoramic image, hotspots will appear with subject titles concerning areas of interest. To learn more, users can simply click on the hotspot. For example, when taking a virtual stroll through the Ames Extremophile Laboratory (Extremophiles are life forms which thrive in harsh conditions on Earth, analogous to those on other planets.), individuals can click on a workstation to learn about an actual microbial sample taken from a salt pond in Baja, California, and how it could provide NASA with insight into early life on Earth and protecting life in space.

Other features include "zoom in" and "pan" buttons for getting closer looks; "next" and "back" buttons that move users through the tour chronologically; color-coded nodes that let users know their current location within a tour; topic and center "badges" that are awarded upon completion of each tour; and an online chat zone for posing questions and discussing the tours with other users.

"Inside NASA" also steps well outside of the NASA field centers' front doors to give its audience an adventurous yet informative panoramic look at some of the heavenly bodies above. These images explore the landscape of the Moon, as seen from the Apollo 17 lunar rover; the Martian surface, as seen from the Viking 2 lander; and the red spot of Jupiter, as seen from the planet's Jovian moon, Europa, via the Galileo orbiter.

Through Seismic Entertainment, Sculley and Brooks have flexed their NASA know-how in developing other interactive educational products. They created the "Exploring Other Worlds" interactive exhibit series for the Smithsonian Institution's National Air and Space Museum. For this effort, Seismic Entertainment was selected for permanent exhibition by the Smithsonian, and won the gold medal at the New York Film Festival in the "Best Museum Exhibit" category.

The company also takes its customers on virtual expeditions of Antarctica and legendary lost cities. Its "Antarctica: Ross Island Area" CD-ROM features panoramas taken by expert researchers and live tours of the National Science Foundation facilities located on the cold, remote continent. Its "Palenque: Lost City of the Maya" CD-ROM employs panoramas and educational material from Dr. Merle Green Robertson, a leading curator of pre-Columbian art at the De Young Museum in San Francisco. A separate-but-similar themed product called "Lost Worlds: Atlas of the Unknown" boasts an interactive electronic atlas of 12 "lost world" regions, with character guides, virtual tours of important cultural, archaeological, and scientific sites within the regions, and live chats with archaeologists.

Seismic Entertainment's interactive software products have been featured on CNN and "Good Morning America," and in Newsweek, Entertainment Weekly, and the San Jose Mercury News. ❖

A virtual visit within NASA's Vehicle Assembly Building, where nose cones for the Space Shuttle solid rocket boosters await assembly. The Vehicle Assembly Building is one of the largest buildings in the world, by volume. With a volume of 3,665,013 cubic meters, it can house nearly four Empire State Buildings. As another indicator of size, the American Flag painted on the exterior of the building required 6,000 gallons of paint, with each stripe on the flag being as big as the tour buses used to transport visitors around Kennedy.

Making a Reliable Actuator Faster and More Affordable

Originating Technology/ NASA Contribution

Before any rocket is allowed to fly and be used for a manned mission, it is first test-fired on a static test stand to verify its flight readiness. NASA's Stennis Space Center provides testing of Space Shuttle Main Engines, rocket propulsion systems, and related components with several test facilities. It has been NASA's test-launch site since 1961.

The testing stations age with time and repeated use; and with aging comes maintenance; and with maintenance comes expense. NASA has been seeking ways to lower the cost of maintaining the stations, and has aided in the development of an improved reliable linear actuator that arrives onsite quickly and costs less money than other actuators.

In general terms, a linear actuator is a servomechanism that supplies a measured amount of energy for the operation of another mechanical system. Accuracy, reliability, and speed of the actuator are critical to performance of the entire system, and these actuators are critical components of the engine test stands.

Stennis Space Center is NASA's lead center for rocket propulsion testing and manages the Agency's rocket propulsion test programs and facilities for current and future space vehicles.

Partnership

An actuator was developed as part of a Dual-Use Cooperative Agreement between BAFCO, Inc., of Warminister, Pennsylvania, and Stennis.

BAFCO identified four suppliers that manufactured actuator components that met the rigorous testing standards imposed by the Space Agency and then modified these components for application on the rocket test stands.

In partnership with BAFCO, the existing commercial products' size and weight were reworked, reducing cost and delivery time. Previously, these parts would cost between $20,000 and $22,000, but with the new process, they now run between $11,000 and $13,000, a substantial savings, considering NASA has already purchased over 120 of the units. Delivery time of the cost-saving actuators has also been cut from over 20 to 22 weeks to within 8 to 10 weeks.

The redesigned actuator is commercially available, and the company is successfully supplying them to customers other than NASA.

Product Outcome

The BAFCO Model 773 is a next-generation, high-thrust, high-response actuator designed to meet performance standards in aerospace, industrial, and petrochemical applications. It meets the National Fluid Power Association standards for cylinder performance, as well as meeting U.S. and European standards for electrical component functions.

The actuators are adaptable for use in a variety of industrial applications, including steam turbines, process-control valves, dampers, for motion control, and a variety of other mechanical purposes. Customers are taking

The high-pressure gaseous oxygen panel which feeds the ignition system for a rocket engine combustion device, located at ground level on the E-I Test Facility, Cell 3. The valve identified as PCV-10A2892-GO is operated by means of the high-performance actuator developed as a part of the Dual-Use Cooperative Agreement with BAFCO, Inc.

advantage of them, since the time and cost of manufacturing are less than those of the other, less-reliable actuators.

Previously used linear actuators worked inconsistently as positioning devices, and their ability to decelerate to gentle stops ranged from unsatisfactory to nonexistent. The BAFCO 773 offers solutions to both of these problems and more.

To solve the problem of positioning reliability, the flow of hydraulic fluid to the two ports of the actuator cylinder is controlled by a servo valve that is controlled by a signal from a servo amplifier that, in turn, receives an analog position-command signal from a supervisory control system of the facility. As the position command changes, the servo valve shifts, causing a greater flow of hydraulic fluid to one side of the cylinder, thereby causing the actuator piston to move to extend or retract a piston rod from the actuator body. A linear variable differential transformer (LVDT), directly linked to the piston, provides a position-feedback signal, which is compared with the position-command signal in the servo amplifier. When the position-feedback and position-command signals match, the servo valve moves to its null position, in which it holds the actuator piston at a steady position.

The actuator also includes a deceleration feature for both extremes of the piston stroke. When the actuator is used to open and close a valve, the deceleration feature prevents damage to valve seats and other components during cycles of rapid stroking. Because the resolution of the LVDT is, for practical purposes, unlimited, the position feedback from the LVDT acts, in conjunction with the deceleration feature, to afford maximum protection against damage in those ranges of position in which protection is most needed.

In addition to answering the previous two problems, the improved actuators help to eliminate leaks associated with common tubing problems by having the components within the actuator connected to a high-pressure hydraulic fluid via a manifold.

The Model 773 has been so successful that BAFCO applied the same concept of off-the-shelf parts modified for specific applications to the development of the Model 775 and 793 as well. The Model 775 rotary actuator was designed as a cost-competitive alternative to pneumatic or electric actuators and operates 90-degree rotation valves of dampers in modulating, on-off and fail-safe applications. Industries that will benefit from this technology include aerospace, petrochemical, well-choke services, power generation, and manufacturing.

The Model 793 linear actuator was designed to extend the existing line of 773 actuators for use with very small linear valves. This new short-stroke actuator featuring auxiliary components common to the Model 773 actuators provides BAFCO the ability to supply actuators to a much larger range of valve sizes.

BAFCO Models 773, 775, and 793 actuators have demonstrated improved accuracy and precision, increased operating speeds, and extended life expectancy over the pneumatic and electric actuators previously used industry-wide. Domestically, BAFCO has made sales to several NASA facilities, Honeywell, the University of Purdue, several U.S. Air Force facilities, and other major suppliers of the aerospace industries. Internationally, BAFCO has sold Model 773 actuators to an oil refinery in Serbia and Model 775 actuators to several oil refineries in China. To date, sales of BAFCO's 773, 775, and 793 lines of actuators, including sales to NASA, total more than $2 million, with no foreseeable end to the market potential. ❖

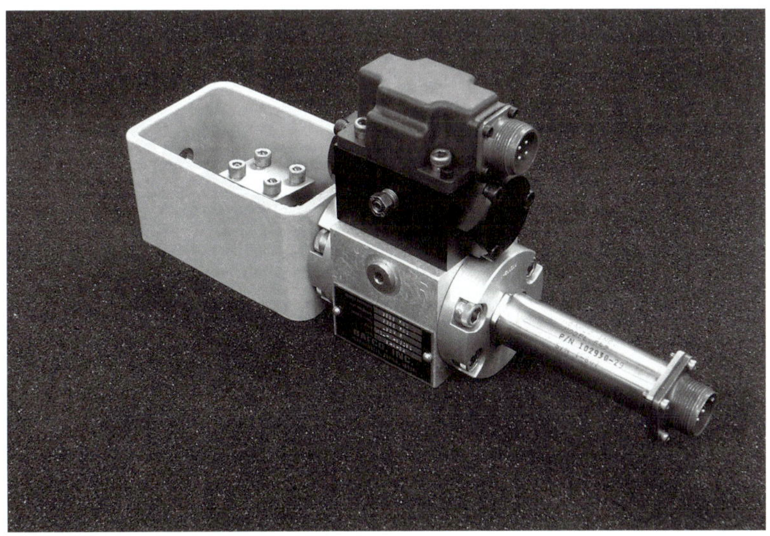

The BAFCO Model 793 linear actuator, one of the newer models in the family of products created in the Dual-Use Cooperative Agreement.

Cost-Cutting Powdered Lubricant

Originating Technology/ NASA Contribution

Scientists at NASA's Glenn Research Center developed a high-temperature, solid lubricant coating material that is saving the manufacturing industry millions of dollars. The material came out of 3 decades of tribological research, work studying high-temperature friction, lubrication, and the wearing of interacting surfaces that are in relative motion. It was developed as a shaft coating deposited by thermal spraying to protect foil air bearings used in oil-free turbomachinery, like gas turbines, and is meant to be part of a larger project: an oil-free aircraft engine capable of operating at high temperatures with increased reliability, lowered weight, reduced maintenance requirements, and increased power.

This advanced coating, PS300, is a self-lubricating bearing material containing chromium oxide, with additions of a low-temperature start up lubricant (silver) and a high-temperature lubricant, making it remarkably stable at high temperatures, and better suited than previously available materials for high-stress conditions. It improves efficiency, lowers friction, reduces emissions, and has been used by NASA in advanced aeropropulsion engines, refrigeration compressors, turbochargers, and hybrid electrical turbogenerators.

PS300 is ideal in any application where lowered weight and reduced maintenance are desired, and high-temperature uses and heavy operating speeds are expected. It has notable uses for the Space Agency, but it has even further-reaching potential for the industrial realm.

Partnership

The Great Lakes Industrial Technology Center (GLITeC), a NASA technology incubator that helps small business take advantage of available NASA technologies, assisted ADMA Products, Inc., in obtaining a license for PS300.

Oil-free bushings coated in PS300 save companies thousands of dollars in repair costs. PS300 is a composite high-temperature, lubricating, chrome-oxide-based material that is embedded with compound particles that function as solid lubricants.

ADMA, based in Hudson, Ohio, specializes in powder metallurgy products from titanium, zirconium, niobium, and other advanced materials and alloys. It had been using PS200 for 8 years already when NASA developed the more advanced PS300. Vladimir Moxson, ADMA president, jumped at this new opportunity. GLITeC worked with company management to design a commercialization plan, and ADMA now holds the license for PS300.

Through GLITeC, NASA continued to support ADMA with the development and commercial application of PS300. Researchers worked closely to optimize the manufacturing process and maximize yield. The result was that, after the collaboration, ADMA was able to increase yields of the key starting material from 5 percent to 45 percent. This improvement reduced the price of the PS300 composite, cut delivery times of the product, and increased ADMA's profits.

ADMA is now supplying the advanced PS300 at a reduced cost to thankful customers, who, in turn, are saving hundreds of thousands of dollars from the use of this space-age product.

Product Outcome

PS300 starts as a powder, which ADMA can either manufacture into a customized solution for a customer's specific needs or apply directly as a coating, via thermal spraying techniques or standard powder metal application methods, such as the press and sinter methods. In short, ADMA can work this material to fit any needs.

One of ADMA's customers, Elliott Turbomachinery Company, of Jeannette, Pennsylvania, has had resounding success with PS304, a derivative of PS300 which is used for coating by plasma spraying. Elliott is a designer and manufacturer of air and gas compressors, steam turbines, power recovery turbines, and power-generating equipment. The company has heavy, moving machinery, called lift rods, that withstand an amazing amount of wear and operate in temperatures up to 1,005 °F.

ADMA provided, through Hohman Plating and Manufacturing Corporation, of Dayton, Ohio, PS300 powdered lubricant-coated valve lift rods for Elliot's steam turbine compressors, with noteworthy economic benefits. Elliott had previously replaced the rods every 2 years, but projects that these new rods will last 8 years. It estimates that this will save at least $3 million in repair costs, not including the additional advantage of not having to leave the equipment idle for days upon days during the actual repairs.

ADMA has provided another company with this dry lubricant and had similarly remarkable cost-saving effects. The Lincoln Electric Company, of Cleveland, Ohio, designs, develops, and manufactures arc welding products, robotic welding systems, and plasma- and oxygen-cutting equipment. Lincoln had the need to replace a set of bronze bushings that were an integral part of a dryer oven used in a manufacturing line. The current bushings were causing system failures, which led to downtime and frequent, costly repairs. ADMA produced a set of PS300 bearings to address the problem.

Lincoln found that these bearings showed such high reliability that the company decided to retrofit all of its sintering furnaces with them. This expenditure, the company estimates, will save an average of $200,000 per furnace per year, over an estimated lifespan of 10 to 20 years. Like Elliot, Lincoln is finding additional economic return, since the furnaces do not have to be shut down as often for maintenance and repairs. It refurbished a number of its furnaces and even its subsidiaries abroad are working to replace their bronze bushings with the space-age material. ❖

This red-hot machine (pictured here with the door open) tests the PS300 bearing materials under very high temperatures.

NASA's Radio Frequency Bolt Monitor: A Lifetime of Spinoffs

Originating Technology/ NASA Contribution

This story begins in the 1970s, when Dr. Joseph Heyman, a young scientist at NASA's Langley Research Center, was asked to support the investigation of a wind tunnel accident at a sister center. Although the work was outside of his physics background, it sparked a research focus that guided his lengthy NASA career and would earn him a slew of accolades, including NASA's highest award medals for Exceptional Leadership, Exceptional Achievement, and Exceptional Service; the coveted "Silver Snoopy" Astronaut Award for Space Shuttle Return to Flight; and the Arthur Fleming Award for being one of the Top Ten Federal Scientists in Government Service. He won 30 additional NASA awards, including the Agency's "Invention of the Year" and the Agency's highest award for technology transfer, and was the only person to ever win 4 R&D 100 Awards.

Deep within a mine shaft, Dr. Joseph Heyman looks on at early testing of NASA's bolt elongation monitor. Heyman is monitoring mine roof bolt preload during a test to compare the P2L2 data to that of a pressure plate. The tests demonstrated that, with the P2L2, each bolt could become its own precision load monitor.

Back in 1973, though, Heyman was a young civil servant with a background in physics who was asked to sit on an accident review panel. The panel met at Ames Research Center, in Moffet Field, California, and after considerable investigation, concluded that a high-pressure pebble heater used for heating gas had failed, due to improperly tightened bolts in a 1,000-pound gate valve control section. The accident showered the facility with incendiary ceramic spheres and nearly a ton of metal, but, luckily, caused no injuries.

Heyman returned to Langley and began work on a solution. He developed an ultrasonic device that would measure bolt elongation, as opposed to torque, the factor typically measured in testing bolt preload or tension. Torque measurement can lead to load errors, with miscalculations as high as 80 percent that can be passed over during installation. Bolt stretch, however, is nearly always accurate to 1 percent or better.

Within 1 month, he had an acoustic resonance solution that accurately determined bolt elongation. He assumed his work on this project had ended, but it was actually the start of nearly 15 years of work perfecting, improving, inventing, and modifying the "bolt monitor"—all the while, filing numerous patents, presenting papers, and holding demonstrations as the technology matured.

Industry engineers challenged Heyman's inventiveness, and reminded the physicist that most bolts are not perfect resonators, and that early devices required that the bolt have reasonably flat and parallel faces. The U.S. Geological Survey asked NASA for help in determining the load in mine roof bolts, which are 8- to 10-feet-long and rough cut. To solve that problem, Heyman modified the original device to operate at a lower frequency and to generate propagation modes that could be used to "lock" the instrument on a particular mode. Further work in this vein led to the development of the Pulsed Phase Locked Loop (P2L2) that worked on the mine bolts.

The next set of problems involved high-strength bolts with head markings. For this solution, Heyman invented a modified P2L2 that tracked a specific phase point in the measurement wave. This class of instrumentation, well suited to measuring small changes in acoustic velocity, won the NASA "Invention of the Year" award in 1982.

Other scientists and engineers have continued the evolution of this technology both inside NASA and outside of the Agency. Within NASA, the technology has been improved for medical applications, with a particular focus on intercranial pressure (ICP) monitoring.

Astronauts exposed to microgravity experience an elevation in ICP, which contributes to space adaptation syndrome, or "space sickness," with symptoms similar to motion sickness on Earth, consisting of general discomfort, queasiness, nausea, vomiting, vertigo, headache, malaise, drowsiness, and lethargy. There is concern for astronauts developing ICP during long-duration missions, because experiencing these symptoms will not allow them to work to full capacity.

A second medical application, unrelated to space travel, involves muscle pressure that can build to a point where blood is unable to properly circulate, and leads to tissue damage and possibly death. This condition is called compartment syndrome. The compartments are groups of muscles surrounded by inelastic fascia, and thus, any swelling of muscles leaves no room for expansion. Blood supply is progressively cut off. Drs. Tom Yost and John Cantrell of Langley are working on applications that relieve the pressure using the P2L2.

Partnership

Heyman retired from the Agency in 2001, and he joined Nascent Technology Solutions, a small Virginia-based company in the field of nondestructive evaluation (NDE) that was recently purchased by Virginia neighbor,

Luna Innovations, Inc. The scientists at Luna use diverse core technologies, including the P2L2 technology developed at Langley; so now, Luna is where Heyman continues his life's work.

He is chief scientist for Luna and working with a new team to build a fully digital P2L2-equivalent device with broad applications ranging from medical to NDE.

Product Outcome

Luna has licensed rights from Langley to continue the work on studying compartment syndrome and ICP as well as found myriad uses for the P2L2 technology in the commercial realm.

Compartment syndrome is a significant cause for surgery among soldiers in Iraq. It stems from blunt trauma causing inflamed muscles, which then creates pressure, resulting in heart strain, as the organ is not getting the blood it needs. The result, if left untreated, may lead to amputation or death. The current treatment involves slicing into the muscle tissue to relieve pressure, which, in addition to being quite excruciating, creates a large wound with risk of infection and a painful recovery time.

The current measurement to determine if the surgery is necessary is a bioptic pressure plunger, another painful procedure involving extraction of tissue from the aggravated area.

With cooperation from NASA's team, Luna's Dr. Ted Lynch has advanced the commercialization of the P2L2 to alleviate the need for this procedure and prevent unneeded surgery. It is already in use in a teaching hospital and should be more widely disseminated as the studies advance.

In addition to its uses in the aforementioned medical applications and its original use of measuring bolts, fasteners, and bond strength—which is still one of the most widespread uses for the P2L2—Luna has found many other innovative applications for this technology.

For example, it anticipates widespread application in the field of radiation dosimetry, the accurate measure of radiation doses for medical purposes. It has teamed with the National Institute of Standards and Technology and demonstrated the ability to measure changes in water temperature of 13 micro-degrees at room temperature. This breakthrough resolution translates into the ability to assess a radiation treatment beam for energy deposited and beam geometry critical for effective treatment. In short, it can make radiation dosages more safe, accurate, and effective.

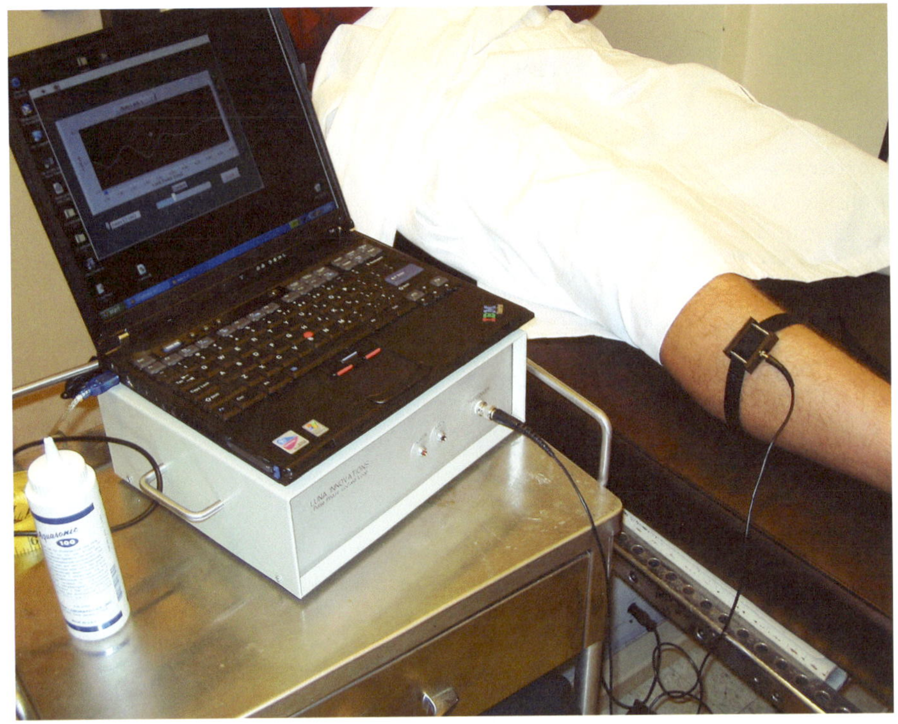

Luna's current compartment syndrome prototype is now being tested for noninvasive diagnosis. A next-generation design will reduce the size of the unit by a fourth and increase functionality. Here, its is shown taking internal pressure reading. The most common method for performing this test is a painful biopsy.

In the field of groundwater analysis, Luna has been able to monitor the slightest flow of groundwater in sand. This work enables a new class of devices capable of determining water flow underground to assess contaminant transport, aquifer impact for pumping, and the movement of oil. This groundbreaking application eliminates the need for breaking too much ground.

Luna has also found a home for the P2L2 in the field of materials characterization. The system is being used by the U.S. Army to measure elastic properties of materials that are related to stiffness as well as nonlinear properties linked to strength. It essentially tells the military how strong things are and how long they will last. This application has uses spread across many additional industries for quality assurance, and life-cycle maintenance.

In the railroad industry, P2L2 is being used as a rail performance-monitoring system. It can detect defects in the rails so that they can be corrected before growing and becoming unsafe.

The P2L2 already has a variety of uses, and the list of applications keeps growing as Heyman continues the work that he was called to do over 30 years ago at Ames. ❖

Going End to End to Deliver High-Speed Data

Originating Technology/ NASA Contribution

By the end of the 1990s, the optical fiber "backbone" of the telecommunication and data-communication networks had evolved from megabits-per-second transmission rates to gigabits-per-second transmission rates. Despite this boom in bandwidth, however, users at the end nodes were still not being reached on a consistent basis. (An end node is any device that does not behave like a router or a managed hub or switch. Examples of end node objects are computers, printers, serial interface processor phones, and unmanaged hubs and switches.)

The primary reason that prevents bandwidth from reaching the end nodes is the complex local network topology that exists between the optical backbone and the end nodes. This complex network topology consists of several layers of routing and switch equipment which introduce potential congestion points and network latency.

By breaking down the complex network topology, a true optical connection can be achieved. Access Optical Networks, Inc., is making this connection a reality with guidance from NASA's nondestructive evaluation experts.

Partnership

Naperville, Illinois-based Access Optical is a fabless semiconductor and optical component subsystem manufacturer. (Fabless refers to a company that does not manufacture its own silicon wafers and concentrates on the design and development of semiconductor chips.) It develops and delivers value-added, high-speed, optical-to-digital transducers that directly connect digital computers to local area networks (LAN) and wide area networks (WAN), as well as metro and long haul networks. These products provide true end-to-end optical connections between server, database host, and LAN domains as high-speed peripheral devices.

Access Optical Networks, Inc.'s end-to-end optical technology not only allows a server to tie directly to an optical area network, but also offloads Internet protocol stack operations. This provides a significant improvement in server performance by freeing up central processing unit (CPU) cycles, reducing system interruptions, and eliminating memory overhead and processor latency.

While searching for opportunities to run proof-of-concept tests on a new router/switch optical memory storage system, Access Optical encountered a team of nondestructive evaluation researchers at NASA's Glenn Research Center. These researchers saw that the high-speed, multi-gigabit and multi-terabit data-transfer capabilities of Access Optical's storage system had potential to enhance a laser-holographic technique they use; this technique involves neural networks to analyze patterns in holographic images. Such data-intensive computation can be a slow process, but the company's technology showed promise in speeding up the analysis and improving output.

An award from the NASA Illinois Commercialization Center—Glenn's link to Illinois businesses—helped initiate the partnership between Access Optical and Glenn, so that the new technology could be proof-tested with help from the nondestructive evaluation team. For Access Optical, this "technology fusion experiment" with NASA provided a specific application on which to focus its efforts, valuable insight into product modifications and product integration, and the financial means to expand as a successful company.

NASA is now looking at Access Optical's high-speed data-processing capability for application in other areas, especially those supporting the Vision for Space Exploration.

Product Outcome

Now a concept turned reality, Access Optical's Alpha-Omega (AΩ) family of optical peripheral equipment resolves data rate mismatches between high-speed optical carrier-grade networks and the digital computer bus interfaces with high-speed, high-density shared memory, and simple bus protocols. According to the company, these resolutions can be achieved at a cost that is 60-percent lower than competitor solutions.

The AΩ optical peripheral units are platform computer interface products with interchangeable network interface

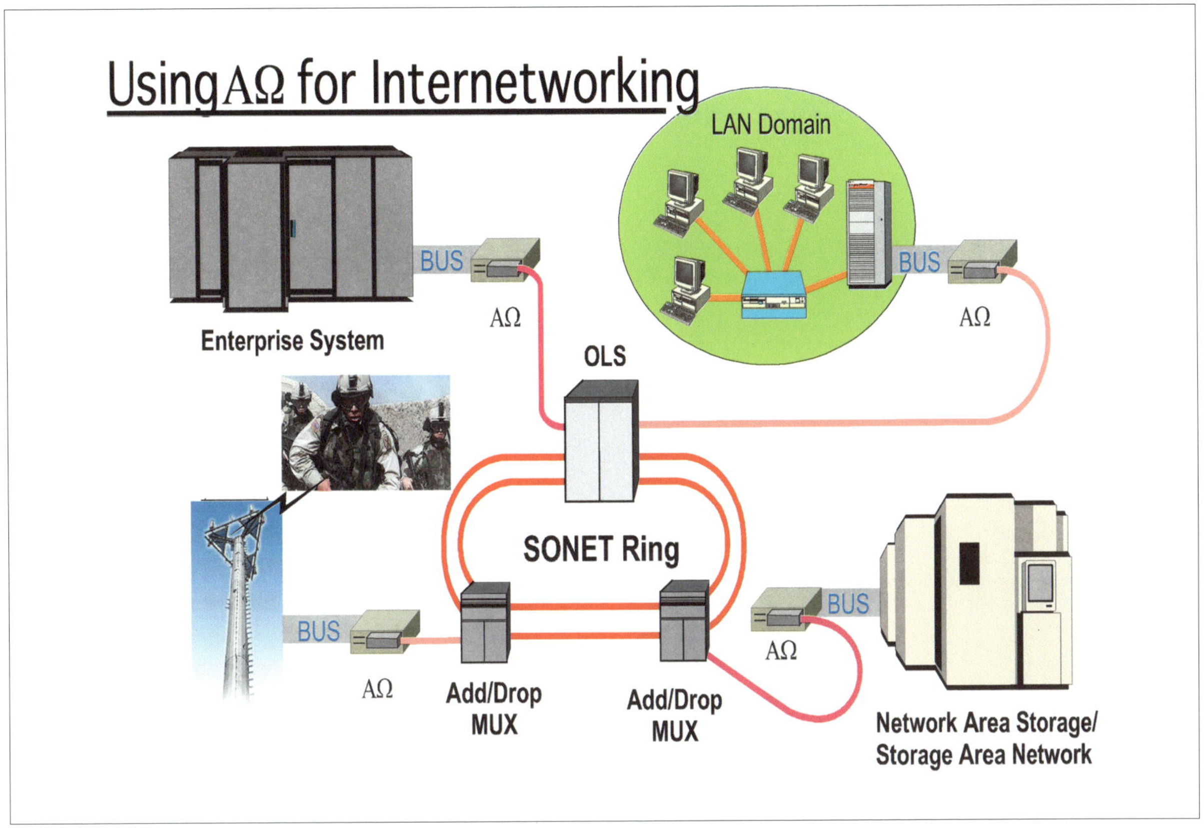

Access Optical Networks, Inc.'s next-generation access point solution.

cards for optical SONET (Synchronous Optical Network Technologies) or optical Ethernet protocol network connections. Currently, corporate servers and computer interfaces rely on disbursement networks, which consist of several layers of routing and switching equipment, to gain access to Ethernet and SONET long haul networks. These disbursement networks add several "hop" junctures and introduce congestion points, thus, diminishing bandwidth. With the AΩ units, bandwidth is increased, disbursement networks are flattened, and congestion points are eliminated.

Access Optical's AΩ products also perform routing functions that allow customers to configure point-to-point and point-to-multipoint transmissions using ring and mesh network topologies. Such configurations translate into peer-to-peer server, workstation, and client/server sub-networks directly connected or seamlessly interconnected via LAN/WAN, metro, and/or long haul networks. Consequently, the devices provide maximum utilization of bandwidth between computers connected via communication networks while eliminating the need for intermediate network equipment (e.g., routers, asynchronous transfer modes, and cross-point switches) to aggregate data traffic and manage intermediate communication protocols. They use a single protocol over a broad range of megabits and terabits that scales to meet the customer's need to change or grow the capacity of their network.

The AΩ technology provides maximum benefits to customers who have high-capacity streaming data applications, such as Wavelength Services, Storage Area Network (SAN)/Network Attached Storage (NAS), and Internet Service Provider (ISP) in the private sector. Other customers can use the products to build high-reliability ring and meshed networks that Access Optical asserts are "ultra secure." The company also notes that its AΩ solutions can result in a 30- to 50-percent reduction in the customer network architecture hierarchy, eliminating multiple vendors, equipment, protocol integration, and reducing operations, administration, maintenance, and provisioning (OAM&P).

Access Optical maintains that the AΩ products will provide protocol scalability that will integrate easily within existing communication networks, while offloading the computing resources vital to applications and mission-critical functions. This is especially important, as all communications—Earth or space—need to have common high-speed systems that are extremely secure and scalable.

The test bed research performed with NASA has additionally led Access Optical to develop a second product line called the Sigma-Epsilon (ΣE) Bridge. The soon-to-be-available technology could extend a LAN's topology to multiple geographic sites through a direct connection to WAN or metropolitan area network (MAN) optical transport networks. By providing a LAN this direct optical connection, bandwidth could be increased and transport network costs could be reduced. ❖

Advanced Joining Technology: Simple, Strong, and Secure

Originating Technology/ NASA Contribution

The space-age materials that NASA employs in its spacecraft and satellites have different attributes than the building materials that can work for Earthly uses. These materials do not behave like the typical construction materials, and therefore, require new methods for construction.

Work done at NASA's Langley Research Center in the realm of active solder joining has led to a new, self-bonding solder that enables high conductivity, as well as the metallic joining of carbon and ceramic materials to a wide range of metals. The original work involved evaluating high- and low-temperature joining technologies for joining carbon composite structures for use in thermal management and reusable launch vehicles. The initial plan for this innovation was to lower the weight of battery packs in satellites. It was a success.

NASA scientists found use for this technology in fabricating a thermal management package for battery compartments in the Earth Observing System (EOS) satellites, but it is also being used by the Agency for space radiator panels. Because it is light, simple to use, and economical, NASA will likely find other uses for this solder, just as outside of the Space Agency, this unique bond is finding many practical applications.

Partnership

Materials Resources International (MRi), of Lansdale, Pennsylvania, teamed with researchers at Langley through a Phase II **Small Business Innovation Research (SBIR)** contract. MRi technologies and products center on materials, systems, and services for joining difficult-to-join materials, dissimilar metals, intermetallics, and ceramics to metals. The joining of such materials is a growing market with application in the electronics, aerospace, power-generation, automotive, chemical-processing, and process machinery industries.

MRi's work with the resulting technology, S-Bond active solder, was so successful, that in 2002, it created a second company devoted just to this work. S-Bond Technologies LLC (SBT) now handles the manufacturing and distribution of this product.

Product Outcome

Commercial development of S-Bond has been progressing for nearly 10 years with emerging applications in thermal management components for electrical packaging, lasers, high-power electronics, ceramics, graphite, carbides, armor, and many other general industrial applications, including the joining of aluminum structures. The main focus of SBT, though, is on electronic packaging and thermal management.

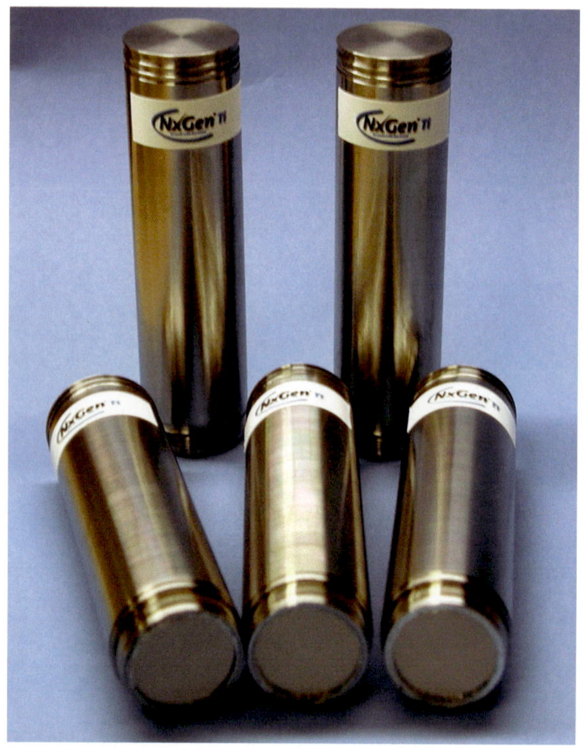

Pictured here are titanium sensor housings for down-hole drilling, with S-Bond-joined ceramic-metal seals on the viewing end of the sensor heads.

Market potential for S-Bond is strongest as a metallic-joining compound that can bond most inorganic materials in an economic, low-temperature, one-step joining process. Market applications include the joining of thermal-management devices that require high-thermal conductivity, the joining of electronic packages that use new ceramics and composite materials, as well as a host of other joining applications where adhesives or conventional solders will not work.

S-Bond works at low temperatures, in air, without flux, and without metallization. It joins all metals, most ceramics, glasses, and metal matrix composites. It can be applied with a variety of methods, including brushing, ultrasonic bath, dipping, thermal spraying, foil or press, and friction transfer, to name a few. Combined with its use of low soldering temperatures, these methods of application make S-Bond very easy to work with. Temperatures for creating the bond can be so low that they can even be achieved with the use of induction, hotplate, or flame. Additionally, S-Bond alloys are lead-free, making them more environmentally acceptable than many traditional solders.

The technology has a growing market demand. From 1997 to 2004, the company has seen sales of over $1 million, and it has emerging electronic packaging markets that hold the potential for sales of over $10 million in the next 3 years.

S-Bond has also been used in space radiators that use carbon composite, graphic foam, and thermal graphite materials for thermal management. In addition, it has been used in circuit board heat transfer panels for cooling of satellites and high electrical density avionics, as well as providing pipe joining for satellite thermal control. It is also available for use in space structures where carbon composite and light metal joining is required. ❖

S-Bond® is a registered trademark of S-Bond Technologies LLC.

Big Results From a Smaller Gearbox

Originating Technology/ NASA Contribution

Many people will be sad to see the Hubble Space Telescope go, as it was the first instrument of its kind to provide us with such a wealth of imagery and information about the galaxy. The telescope has served us well since its launch in spring of 1990, but it is nearly time for its retirement. The science, however, will continue, as NASA plans the launch of a new, more modern orbiting telescope, the James Webb Space Telescope.

Named after the man who ran NASA from 1961 to 1968—years fraught with the anxiety and uncertainty of the Space Race—the scope is scheduled for launch in fall of 2011. It is designed to study the earliest galaxies and some of the first stars formed after the Big Bang.

NASA scientists at the Goddard Space Flight Center are busy developing the technologies to build this new machine. Many of the new technologies are available for commercial licensing and development. For example, the NASA Planetary Gear System technology developed to give precise nanometer positioning capabilities for the James Webb Space Telescope is now being employed by Turnkey Design Services, LLC (TDS), of Blue Island, Illinois, to improve electric motors.

This revolutionary piece of technology allows more efficient operation of the motors, and is more cost-effective than traditional gearbox designs.

Partnership

TDS is located just outside of Chicago, and is a small, full-service product design and analysis firm that specializes in hydraulic, pneumatic, and turbomachinery components and systems. The engineers at this company have experience working for companies in the aerospace, automotive, medical, and consumer product industries. One of the focuses of their design and manufacturing is high-tech gearboxes and their components.

The NASA Illinois Commercialization Center (NICC) provided TDS with a NICC Commercialization Award to license James Webb Space Telescope gear-bearing technologies for development of commercial applications in collaboration with Goddard.

The gear-bearing technology was a NASA industrial breakthrough that improved design by adding strength and improving performance while managing to reduce size, parts-count, and cost.

Product Outcome

Robert Kennedy, president and owner of TDS, says of this project, "We are excited about this technology and look forward to reducing conventional gear set cost and weight while increasing product reliability." The revolutionary gear-bearing technology combines gears and bearing in one package, which provides a host of benefits.

The gear-bearing planetary transmission provides more even gear distribution loading and higher gear ratio in smaller packages. Use of the gear-bearing technology can increase the reliability of electric and hydraulic motor driven actuators, pumps, compressors, and gearboxes. TDS is using this technology as a planetary speed reducer for electric motors that currently use plastic, brass, or steel gears.

TDS is marketing three different prototype sets, each with the same diametral and axial envelopes, but with different ratios. The company sees a wide spectrum of uses for these sets, but intends to market them to the aerospace and automotive industries. The planetary sets could be used in the aerospace industry to open and close bay doors of aircraft and to extend and retract aircraft wing flaps. In the automotive industry, this technology could be used to open and close sliding doors on vans, to open and close lift gates on sport utility vehicles, and to move powered seats back and forth. ❖

Although the James Webb Space Telescope is still under development and not charted to fly until 2011, it is already providing us with new science, like this gear system.

Low-Pressure Generator Makes Cleanrooms Cleaner

Originating Technology/ NASA Contribution

Scientists at NASA's Kennedy Space Center work in cleanrooms: laboratories with high degrees of cleanliness provided by strict control of particles such as dust, lint, or human skin. They are contaminant-free facilities, where the air is repeatedly filtered, and surfaces are smooth to prevent particles from getting lodged. Technicians working in these environments wear specially designed cleanroom "bunny suits" and booties over their street clothes, as well as gloves and face masks to avoid any contamination that may be imparted from the outside world. Even normal paper is not allowed in cleanrooms—only cleanroom low-particulate paper. These are sensitive environments where precision work, like the production of silicon chips or hard disk drives, is performed.

Often in cleanrooms, positive air pressure is used to force particles outside of the isolated area. The air pressure in the Kennedy cleanrooms is monitored using high-accuracy, low-differential pressure transducers that require periodic calibration. Calibration of the transducers is a tricky business. In prior years, the analysis was performed by sending the transducers to the Kennedy Standards Laboratory, where a very expensive cross-floated, labor-intensive, dead-weight test was conducted.

In the early 1990s, scientists at Kennedy determined to develop a technique and find equipment to perform qualification testing on new low-differential pressure transducers in an accurate, cost-effective manner onsite, without requiring an environmentally controlled room. They decided to use the highly accurate, cost-effective Setra Model C264 as the test transducer.

For qualification testing of the Setra, though, a portable, lower-cost calibrator was needed that could control the differential pressure to a high degree of resolution and transfer the accuracy of the Standards Laboratory testing to the qualification testing. The researchers decided that, to generate the low-differential pressure setpoints needed for qualification testing, very small gas volume changes could be made against the test article, and a corresponding pressure change would be detected by a pressure standard. This allowed the researchers to recreate cleanroom air pressure settings without the use of a cleanroom.

Thus was born the low-differential pressure generator. In 1993, a prototype was developed using a pair of PVC tanks, a volume controller, and a 1-pound-per-square-inch pressure standard. By 1995, the prototype was perfected into the unit that is still used today.

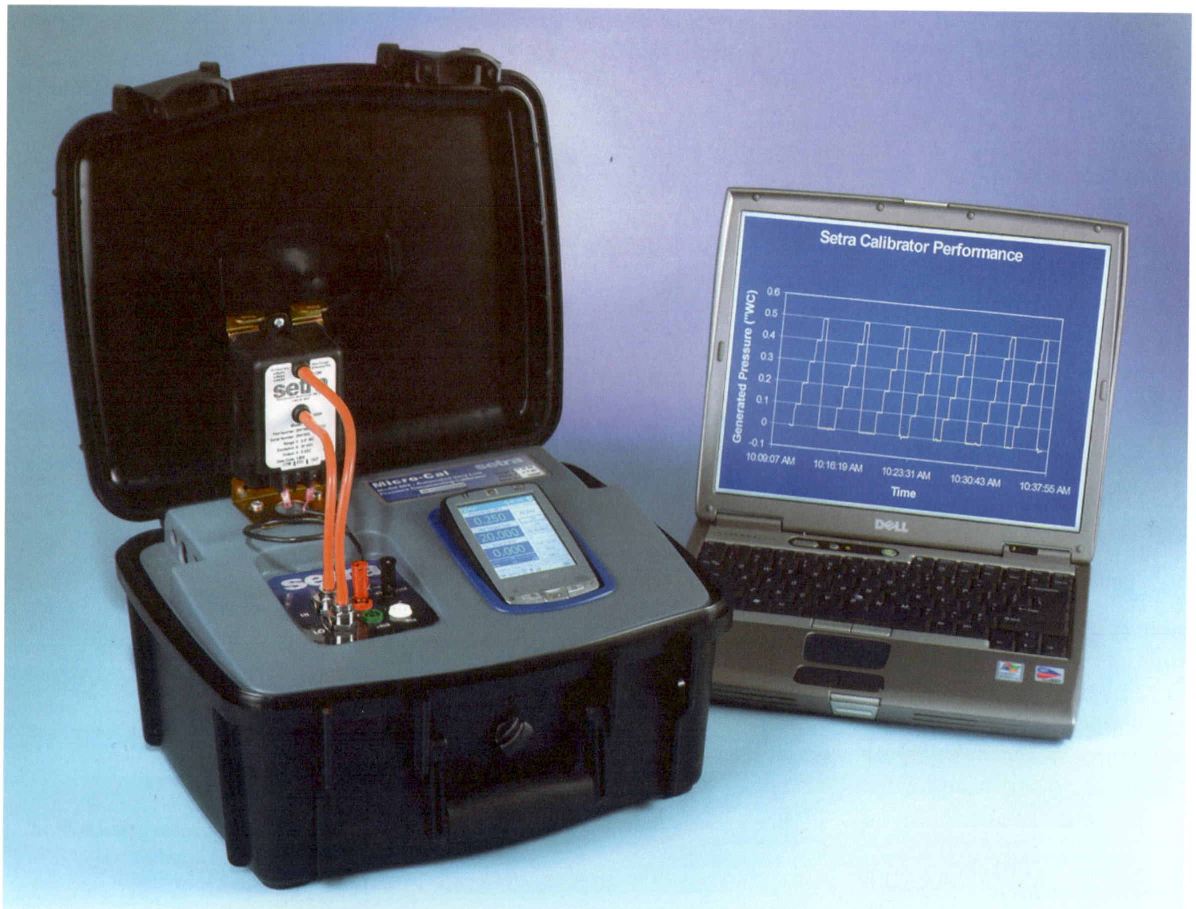

Setra Systems, Inc., has produced the first portable, low-pressure calibration system capable of performing in situ calibrations of high accuracy, very low-differential pressure transducers.

Partnership

As with so many NASA-inspired inventions, the scientists were in need of a new piece of equipment, so they built it themselves. Stephen Stout and Richard Deyoe of Kennedy were the two principal researchers on this project and they, with Greg Hall, patented it as the Low Differential Pressure Generator in 1997.

Ironically, personnel at Setra Systems, Inc., the Boxborough, Massachusetts company whose original Model C264 pressure transducer had been tested, came across the new technology in 2002, while conducting a search for the same type of equipment that the staff at Kennedy had wanted. The new low-pressure generator was described in an article in NASA Tech Briefs. Setra was seeking a pressure-generation method that would isolate the differential pressure sensors from environmental noise during the calibration procedure, a problem that was discovered while working with pharmaceutical manufacturers attempting to certify critical air handling processes.

The unit is battery powered, with compact size and a lightweight case that can be carried to cramped, remote locations, even up stairs or ladders.

Setra then contacted Kennedy's Technology Transfer Office to obtain rights to the patent, and the NASA office facilitated the paperwork and provided them exclusive rights to the technology.

Product Outcome

Setra, known for its simplicity of design, high accuracy, exceptional long-term stability, and competitive pricing, incorporated the NASA technology into its Micro-Cal Low Pressure Calibrator, and now offers this unit among its product line.

Technicians have just one portable unit, instead of having multiple components, like the pressure indicator, pressure generator, and data logger to carry to calibration sites. The unit is battery powered, with compact size and a lightweight case that can be carried to cramped, remote locations, even up stairs or ladders. It is a significant improvement over the expensive primary pressure standards that the transducers had to be tested against. Additionally, it boasts superb pressure-reading accuracy, as well as fast, stable, repeatable, and accurate pressure generation. The user can also select or configure pressure generation profiles.

It offers calibration data storage with download capability, data and process security, dual reference pressure sensors to cover a wide range of test pressures at the highest accuracy possible, and calibration times as fast as 5 minutes per unit, which generate significant laboratory cost savings. It also has a simple personal digital assistant user interface. Most attractive, even with all of these added features, it is almost half the cost of a laboratory benchtop calibrator.

Most manufacturers of low-pressure calibrators do not use true low-range reference pressure sensors. Instead, they use higher-range sensors and attempt to achieve high accuracy at lower pressures through intricate microprocessor correction. The resulting higher levels of noise and instability limit the ultimate accuracy available. Setra uses patented stretched diaphragm, capacitive sensor technology for highest output at the lowest pressures.

Another area where typical low-pressure calibrators fall short is generating a stable, repetitive, and accurate test pressure. Most companies use micro-solenoid pressure generation and regulation, a technique that applies small pressure pulses to the positive and negative pressure test volumes to regulate the test pressure. During active pressure regulation, this system generates pneumatic noise. Setra uses NASA's low-differential pressure-generation technology that produces maximum pressure-setting sensitivity with minimum noise.

The pressure generation is accomplished using a piston/cylinder arrangement, whereby the differential pressure sensor under test has both high- and low-pressure ports connected to the cylinder in a push/pull configuration. As the stepper motor-driven piston advances in the cylinder, it applies positive pressure to the high port of the test pressure sensor and negative pressure to the low port. The resulting pressure-generation system is sealed and immune to the outside environmental noise and has twice the sensitivity as a single-sided piston and cylinder.

Setra has automated the patented NASA Low Pressure Generator using micro-stepping motors and true low-pressure reference transducer feedback. This combination has produced the first portable, low-pressure calibration system capable of performing in situ calibrations of high accuracy, very low-differential pressure transducers. ❖

Micro-Cal™ is a trademark of Setra Systems, Inc.

The Space Laser Business Model

Originating Technology/ NASA Contribution

Creating long-duration, high-powered lasers, for satellites, that can withstand the type of optical misalignment and damage dished out by the unforgiving environment of space, is work that is unique to NASA. It is complicated, specific work, where each step forward is into uncharted territory.

In the 1990s, as this technology was first being created, NASA gave free reign to a group of "laser jocks" to develop their own business model and supply the Space Agency with the technology it needed. It was still to be a part of NASA as a division of Goddard Space Flight Center, but would operate independently out of a remote office.

The idea for this satellite laboratory was based on the Skunk Works concept at Lockheed Martin Corporation. Formerly known as the Lockheed Corporation, in 1943, the aerospace firm, realizing that the type of advanced research it needed done could not be performed within the confines of a larger company, allowed a group of researchers and engineers to essentially run their own microbusiness without the corporate oversight. The Skunk Works project, in Burbank, California, produced America's first jet fighter, the world's most successful spy plane (U-2), the first 3-times-the-speed-of-sound surveillance aircraft, and the F-117A Nighthawk Stealth Fighter. Boeing followed suit with its Phantom Works, an advanced research and development branch of the company that operates independent of the larger unit and is responsible for a great deal of its most cutting-edge research.

NASA's version of this advanced business model was the Space Lidar Technology Center (SLTC), just south of Goddard, in College Park, Maryland. Established in 1998 under a Cooperative Agreement between Goddard and the University of Maryland's A. James Clark School of Engineering, it was a high-tech laser shop where a small group of specialists, never more than 20 employees, worked all hours of the day and night to create the cutting-edge technology the Agency required of them.

Drs. Robert Afzal and Joseph Dallas were directors of the SLTC, and led the development and production of active spaceborne, remote-sensing, optical instruments. As a pioneer in the area of photonics, Dr. Dallas led basic research, development, and production of semiconductor laser diode products, improving coupling efficiency through novel physical optics modeling and intracavity phase-correction techniques. He worked for NASA for 15 years, 11 of which were as a civil servant, and 4 of which were as a contractor.

In the years at the SLTC, Dr. Dallas and the team delivered flight-ready lasers to NASA. The Geoscience Laser Altimeter System (GLAS) space flight lasers, for example, were designed, developed, and assembled at the SLTC. These lasers were integrated at Goddard as part of the Ice, Cloud, and Land Elevation Satellite (ICESat), NASA's benchmark Earth Observing System mission for measuring ice sheet mass balance, cloud and aerosol heights, as well as land topography and vegetation characteristics. Subsequently, the Mercury Laser Altimeter laser was similarly developed and assembled at the facility and left Earth on August 3, 2004, aboard the MErcury Surface, Space ENvironment, GEochemistry, and Ranging (MESSENGER) spacecraft heading to Mercury to measure the topography of the northern hemisphere.

The photonics work performed at the SLTC was on the forefront of a relatively immature branch of scientific study. Contrary to electronics work where engineers have decades of advanced research and flight qualified experience to draw upon, the understanding of issues facing long-duration operation of lasers in space is in its infancy.

Partnership

Drs. Afzal and Dallas and their team of experts, consisting of a specialist in each discipline, managed to do the type of work that could not be performed within a large firm. Eventually, the group disbanded. People left. Programs changed. The amount of work NASA supplied was not steady enough to keep this band of specialists busy or to make the SLTC a viable long-term project. In 2003, Dr. Dallas used the knowledge gained at NASA to found a company, Avo Photonics, in Horsham, Pennsylvania.

Product Outcome

Dr. Dallas learned how a tech company should run from the unique opportunity of working with a team of specialists performing cutting-edge photonics work. He assembled a new team of hard-working experts to drive Avo Photonics.

Because he was supplying more than one customer, NASA, his team would have enough projects to keep themselves busy, while also still being able to promise the same quality of work to its regular customers. And when NASA needed a laser team in the future, it would be able to rely on Avo Photonics, rather than have to assemble a new team every time the need arose.

An aerospace and military center, for example, might need a new photonic instrument every 7 years. It can turn to the same provider each time, a group that knows the

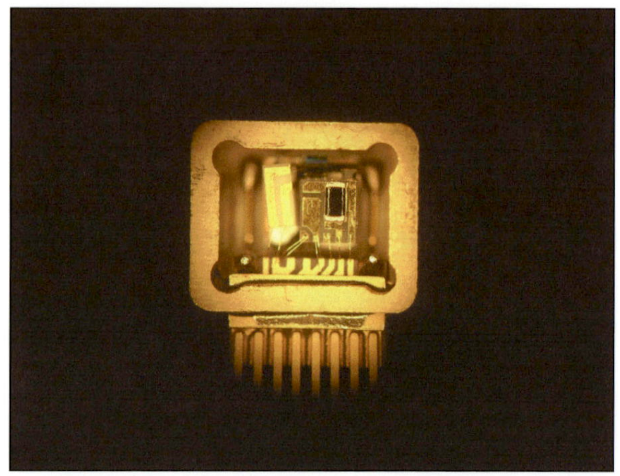

Laser packaging manufactured by Avo Photonics.

Geoscience Laser Altimeter System (GLAS) testing at NASA's Space Lidar Technology Center, where the founders of Avo Photonics "cut their teeth."

specific needs of the customer and has worked on their projects before. And, because it is only after 7 years that it requires a new laser, it need not maintain its own laser shop. Avo Photonics can then rotate business to supply its other customers.

Avo Photonics provides advanced contract manufacturing services for photonic and microelectronic customers in the communications, military, aerospace, medical, and industrial markets. With its state-of-the-art equipment and design expertise for optical and radio frequency products, Avo Photonics is able to support product development from concept through production. Prototyping through volume manufacturing, electro-optic performance testing, high-value design, qualification testing for ruggedness and reliability, fixture development, and lower-cost manufacturing are all part of the company's standard offerings.

Avo Photonics emphasizes a system engineering approach toward the development of solutions for its customers. The strong optical, mechanical, electrical, thermal, and materials engineering design, modeling, and analysis work is continuously guided by the system needs of performance, quality, cost, and schedule. The company understands its customer and meets them at their current level of product design: from fundamental component laser physics to demonstrated product. The engineering and science tools are utilized to move toward cost-effective product realization. This could include aerospace lasers, deployed sensor systems, volume communication lasers, modulators, detectors, industrial ranging systems, or commercial imagers.

Avo Photonics is unique in its combination of high-level engineering support and prototype-to-production facility capabilities. Its customers have a one-stop-shop to see their visions realized. With over 6,000 square feet of cleanroom space and high-resolution tools such as automated component placement machines, wire bonders, laser welders, lid sealers, optical fiber attach stations, lasers, high-accuracy optical mounts, 20-gigabyte network analyzers, precision cleaning equipment, and optical inspection tools, Avo Photonics is fully capable of seeing a design through to production.

Understanding and testing designs, processes, and products for various environments is critical to realization of active and passive photonic systems. Avo Photonics has in-house equipment for thermal cycling, high-humidity exposure, hot/cold storage, optical inspection and measurement, gross and fine leak checking, and shear and pull mechanical strength test, along with numerous standard photonic tools such as power meters, spectrum analyzers, thermal sensors, and electrical drivers all customized into LabVIEW-controlled test stations.

Customers of Avo Photonics see the company as an extension of their own engineering and production team. The company takes great care in every aspect of product development, from finite element analysis through supply chain management, to detailed reporting and failure analysis. The philosophies realized and nurtured at NASA continue to guide the quality and pride of Avo Photonics. ❖

With over 6,000 square feet of cleanroom space and high-resolution tools such as automated component placement machines, wire bonders, laser welders, lid sealers, optical fiber attach stations, lasers, high-accuracy optical mounts, 20-gigabyte network analyzers, precision cleaning equipment, and optical inspection tools, Avo Photonics is fully capable of seeing a design through to production.

RESEARCH AND DEVELOPMENT

AT NASA

The success of the Vision for Space Exploration relies upon the ongoing research and development activities conducted at each of NASA's 10 field centers. In an effort to promote synergy across NASA as it works to meet its long-term goals, the Agency formed four Mission Directorates: Space Operations, Exploration Systems, Science, and Aeronautics Research. These directorates provide NASA Headquarters and the field centers with a streamlined approach to continue exploration in space and on Earth.

Research and Development at NASA

Space Operations

NASA's Space Operations Mission Directorate provides many critical enabling capabilities that make possible much of NASA's science, research, and exploration achievements. It does this through the three themes of the Space Shuttle Program, the International Space Station (ISS), and Flight Support:

- The Space Shuttle Program builds on the Shuttle's primacy as the world's most versatile launch system. The Space Shuttle, first launched in 1981, returned to flight in 2005, with Discovery carrying the STS-114 crew to the ISS.
- The ISS establishes a permanent human presence in Earth orbit. It also provides a long-duration, habitable laboratory for science and research activities investigating the limits of human performance, expanding human experience in living and working in space, and enabling the commercial development of space.
- Flight Support consists of Launch Services, Space Communications, and Rocket Propulsion Testing. These "enabling" services are critical for conducting space exploration, as well as aeronautical, materials science, biological, and physical research.

From left, STS-114 astronauts Stephen Robinson, James Kelly, Andrew Thomas, Wendy Lawrence, Charles Camarda, Eileen Collins, and Soichi Noguchi.

Humans in space are the primary focus of this directorate. Space is still the new frontier, and astronauts are the pioneers of that frontier. The directorate explains, explores, and chronicles the space projects humans are involved in now and will be involved in, come the future.

Space Shuttle: Return to Flight

Space Shuttle Discovery launched from Kennedy Space Center on July 26, 2005, ending a 2.5-year wait for the historic Return to Flight mission. STS-114 included breathtaking in-orbit maneuvers, tests of new equipment and procedures, a first-of-its-kind spacewalking repair task, and telephone calls from two world leaders.

Discovery touched down on August 9 at Edwards Air Force Base, California, following a successful reentry. The orbiter returned to Kennedy on August 21, atop a modified Boeing 747 called the Shuttle Carrier Aircraft. Discovery then entered the Orbiter Processing Facility, where it will be readied for mission STS-121.

During STS-114, NASA accomplished a variety of goals while also learning some important lessons. At liftoff, a large piece of insulating foam broke off the External Tank. Now, NASA engineers are working to determine what caused this and how to prevent it from happening in the future.

Using the new Orbiter Boom Sensor System, Discovery crewmembers took an unprecedented up-close look at the orbiter's Thermal Protection System. This

On July 26, 2005, Space Shuttle Discovery launched into a clear blue sky on the historic Return to Flight mission, STS-114.

collection of new data was expanded on flight day 3, when Commander Eileen Collins guided Discovery through the first-ever "rendezvous pitch maneuver" as the orbiter approached the ISS for docking.

The slow-motion backflip allowed Space Station crewmembers John Phillips and Sergei Krikalev to snap high-resolution photographs for mission managers to use to ensure Discovery was in good shape to come home.

Space Shuttle Discovery is docked to the International Space Station's Destiny laboratory, with the Earth's horizon in the background.

During the first of three spacewalks, Mission Specialists Stephen Robinson and Soichi Noguchi tested new repair techniques for the outer skin of the Space Shuttle's heat shield and installed equipment outside the ISS. They also repaired a control moment gyroscope. Two days later, Robinson and Noguchi again ventured out into the vacuum of space to replace a different, failed control moment gyroscope, putting all four of the Station's gyroscopes back into service.

When two thermal protection tile gap-fillers were spotted jutting out of Discovery's underside, astronauts and other experts on the ground devised a plan to ensure that the protrusions would not cause higher-than-normal temperatures on the Space Shuttle during atmospheric reentry.

Ground controllers sent up plans to the Shuttle-Station complex for Robinson to ride the Space Station's robotic arm beneath the Shuttle and, with surgical precision, pluck out the gap-fillers.

Work on the Shuttle underbelly had never been tried before, but with Mission Specialist Wendy Lawrence and Pilot Jim Kelly operating the robotic arms, Mission Specialist Andy Thomas coordinating, and fellow spacewalker Noguchi keeping watch, Robinson delicately completed the extraction during the third and final spacewalk.

"Okay, that came out very easily," Robinson said, after carefully removing one of the fillers. "It looks like this big patient is cured."

The crew received phone calls from U.S. President George W. Bush and Japanese Prime Minister Junichiro Koizumi, who offered congratulations and appreciation for the astronauts' hard work.

Stephen Robinson is attached to a foot restraint on the International Space Station's Canadarm2. This robotic extension guided Robinson to the underside of Discovery, where he removed two pieces of ceramic fabric, known as "gap fillers," that were protruding from heat-shielding tiles.

A close-up view of Discovery's underside is featured in this image photographed by Robinson—whose shadow is visible on the thermal protection tiles—during the mission's third session of extravehicular activities.

Together, both the Discovery and ISS crews paid tribute to the astronauts of Columbia, as well as others who gave their lives for space exploration.

With the mission drawing to a close, the Multi-Purpose Logistics Module, Raffaello, was removed from the ISS and reinstalled in Discovery's payload bay. Raffaello arrived with more than 12,000 pounds of equipment and supplies and carried about 7,000 pounds of Station material on the trip back to Earth. After 9 days of cooperative work, Discovery undocked from the ISS.

The STS-114 crew was given an extra day in orbit on August 8, when the first attempt to land at Kennedy was foiled by uncooperative weather. Even though cloudy skies reappeared at the Shuttle's home port the next morning, NASA was ready with a backup plan: a landing at Edwards Air Force Base in the high desert of California, where the weather was perfect.

Capsule Communicator Ken Ham congratulated the returning crew on a spectacular test flight. "Stevie Ray, Soichi, Andy, Vegas, Charlie, Wendy, and Eileen—welcome home, friends."

Those words, Collins said, were great to hear. "We're happy to be back, and we congratulate the whole team for a job well done."

The Sun rises on Discovery as it rests on the runway at Edwards Air Force Base, California, after a safe landing August 9, 2005, to complete the STS-114 mission.

International Space Station: Sustaining a Human Presence in Space

While awaiting Discovery's arrival, Expedition 11 NASA Science Officer John Phillips and Commander Sergei Krikalev conducted the first of their three renal stone experiment sessions aboard the ISS. The renal stone experiment investigates whether potassium citrate, a proven Earth-based therapy used to minimize renal (kidney) stone development, can be effective as a countermeasure to reduce the risk of kidney stone formation for crewmembers in space. Astronauts are at an increased risk of developing kidney stones, because urinary calcium levels are typically much higher in space.

The renal stone investigation was designed as a double-blind study. The crewmembers do not know whether they are taking the potassium citrate or a placebo. Further, the principal investigator who interprets the data does not know in advance which crewmembers have taken the potassium citrate or which have taken the placebo. The principal investigator is studying the urine chemistry of the samples to determine each individual's risk of renal stone formation. If the investigator's hypothesis is correct, the crewmembers identified as having a lower renal stone formation risk will be those who had taken the potassium citrate pills in-flight as a countermeasure.

During their initial session, Phillips and Krikalev performed a urine collection over the course of 24 hours and logged everything they ate and drank for 48 hours.

This experiment is crucial to long-duration missions, since kidney stones can incapacitate a crewmember, and, in the worse case, threaten life if there is no way to get the astronaut back to Earth quickly.

In a previous ISS research effort, Expedition 10 Commander and NASA Science Officer Leroy Chiao and Flight Engineer Salizhan Sharipov conducted an experiment to shed more light on what is currently known about microgravity's effects on human muscle and bone.

The International Space Station was photographed from Space Shuttle Discovery after the two spacecraft undocked.

In carrying out the ADvanced Ultrasound in Microgravity (ADUM) experiment, Chiao and Sharipov performed ultrasound bone scans on each other by taking turns as operator and subject. The bone scans were taken of the shoulder, elbow, knee, and ankle, monitored remotely from the ground, and videotaped and photographed for downlink and analysis.

Since there is no room for a fully functioning staff of doctors aboard the ISS, nor is it feasible for a crewmember to return to Earth for a quick medical checkup, this experiment could lead to efficient diagnosing of medical problems with minimal use of onboard resources. Ability of crewmembers to use an ultrasound machine with remote instruction—sending information to the ground for analysis—can assist in timely treatment, as well as avert unnecessary evacuation. Crewmembers as far away as Mars could eventually be remotely examined by doctors on Earth using a modification of this technology. This type of capability is essential for long-term space exploration.

The Expedition 10 crewmembers also conducted a session with the Miscible Fluids in Microgravity experiment. Fluids do not behave the same on Earth as in the microgravity environment inside the orbiting Space Station. This experiment studies how miscible fluids, or those that completely dissolve, interact without the interference of gravity.

The test involved Chiao pulling tinted water from a syringe through a drinking straw and into another syringe containing a mixture of honey and water. The way the fluid interacted was videotaped and photographed for observation. This research could help scientists improve the way plastics and other polymers are produced on Earth and in space.

NASA's Payload Operations team at Marshall Space Flight Center is coordinating the aforementioned ISS science activities.

Flight Support: Launch Services

Many of NASA's most famous missions are launched on expendable launch vehicles (ELVs). These missions are unpiloted and can accommodate all types of orbit inclinations and altitudes/attitudes.

In 1997, Kennedy was assigned lead center program responsibility for NASA's acquisition and management of ELV launch services. Its ELV Program Office provides a single focal point for these services, while affording NASA the benefits of consolidated and streamlined technical and administrative functions. The program, with its vision statement, "Global Leadership in Launch Service Excellence," provides launch services for NASA, NASA-sponsored payloads, and other government payloads.

Primary launch sites are Cape Canaveral Air Force Station, Florida, and Vandenberg Air Force Base, California; other launch locations are NASA's Wallops Island, Virginia; Kodiak Island, Alaska; and Kwajalein Atoll, in the Republic of the Marshall Islands, in the North Pacific.

Since 1990, NASA has been purchasing ELV launch services directly from commercial providers, whenever possible, for its scientific and applications missions that are not assigned to fly on the Space Shuttle. Because ELVs can accommodate all types of orbit inclinations and altitudes/attitudes, they are ideal for launching Earth-orbit and interplanetary missions.

Many of NASA's most famous missions—from those observing Earth, such as EOS, Aura, and Landsat, to interplanetary and deep space missions like the Mars Exploration Rover and Deep Space 1—are launched on expendable launch vehicles.

Kennedy is also responsible for NASA oversight of launch operations and countdown management. A motivated and skillful team is in place to meet the mission of the ELV program: "To provide launch service excellence, expertise, and leadership to ensure mission success for every customer."

In late-May 2005, NASA successfully launched a new environmental satellite for the National Oceanic and Atmospheric Administration (NOAA), using a Boeing Delta II 7320-10 ELV. The satellite, NOAA-18, aims to improve weather forecasting and monitor environmental events around the world.

The NOAA-18 spacecraft lifted off from Vandenberg Air Force Base, on the Delta II. Approximately 65 minutes later, the spacecraft separated from the ELV second stage.

"The satellite is in orbit and all indications are that we have a healthy spacecraft," said Karen Halterman, the NASA Polar-orbiting Operational Environmental Satellites (POES) project manager, based at Goddard

NOAA-18 is the latest polar-orbiting satellite developed by NASA for the National Oceanic and Atmospheric Administration (NOAA). NOAA-18 will collect information about Earth's atmosphere and environment to improve weather prediction and climate research across the globe.

The National Oceanic and Atmospheric Administration's spacecraft NOAA-18 leaped away from the smoke and steam clouds as it lifted off from Vandenberg Air Force Base in California. It was launched by NASA on a Boeing Delta II 7320-10 expendable launch vehicle.

Space Flight Center. "NASA is proud of our partnership with NOAA in continuing this vital environmental mission," she added.

NOAA-18 will collect data about the Earth's surface and atmosphere. The data are input to NOAA's long-range climate and seasonal outlooks, including forecasts for El Niño and La Niña. NOAA-18 is the fourth in a series of five POES with instruments that provide improved imaging and sounding capabilities.

NOAA-18 has instruments used in the International Search and Rescue Satellite-Aided Tracking System, called COSPAS-SARSAT, which was established in 1982. NOAA POES detect emergency beacon distress signals and relay their location to ground stations, so rescue can be dispatched. SARSAT is credited with saving approximately 5,000 lives in the United States and more than 18,000 worldwide.

NOAA manages the POES program and establishes requirements, provides all funding, and distributes environmental satellite data for the United States. Goddard procures and manages the development and launch of the satellites for NOAA on a cost-reimbursable basis.

Flight Support: Space Communications

Sophisticated signal-processing techniques and simple proof-of-principle antenna arrays built from PVC pipe, aluminum foil, and copper wire could revolutionize the way NASA obtains data from its Earth-observing satellites.

If the adaptive array system being studied by NASA and Georgia Institute of Technology (Georgia Tech) researchers ultimately proves feasible, it could make information from the Space Agency's Earth-observing satellites more widely and rapidly available. The "off-the-shelf" technology has already demonstrated that it can successfully receive one satellite telemetry frequency.

"The dream would be to make these NASA information services available to anybody sitting at a computer, almost like video-on-demand," explained Mary Ann Ingram, a professor at Georgia Tech's School of Electrical and Computer Engineering. "Timely information from Earth-observing satellites could be useful in many ways, such as directing operations to fight a forest fire, for instance."

Information from satellites such as Earth Observing-1 (EO-1) is now downlinked to various 11-meter dishes, primarily in the Arctic Circle, where subzero temperatures create maintenance and reliability issues for their complex aiming mechanisms. Typically, satellites such as EO-1 are in contact with these antenna systems 5 to 8 times a day, for 10 minutes at a time. The present antenna systems require resident crews to operate and maintain them.

The NASA/Georgia Tech project envisions replacing these antennas with a network of inexpensive antenna arrays that would have no moving parts and use sophisticated software—instead of careful aiming—to gather data from the satellites. The network could lower operational costs while improving access to the information.

"When people use cell phones to make calls, there are no moving parts on the antennas," noted Dan Mandl, mission director for NASA's EO-1 program at Goddard. "What I would like to do is build a continuous cell-like network around the world that would provide almost unlimited opportunities to downlink data."

Mandl compared NASA's existing downlink system to old-fashioned pay phones located off expressway exits. "If you witness an accident, you can open your cell phone and call for assistance," he said. "But if you don't have a cell phone, you have to get off the highway at the next exit and hunt for a pay phone. What we would like to do is give these satellites the equivalent of cell phones to allow anytime, anywhere contact."

The proof-of-principle adaptive arrays being tested by Ingram and her research team are built from inexpensive components, including common PVC piping and aluminum foil. Signals from the four antennas are analyzed using a processing technique that learns to improve its performance, by constructively combining scattered and reflected versions of the signal and by suppressing noise and interference. This eliminates the need for costly front-end hardware and precise aiming of the antenna arrays, and enables flexibility in the location of the ground station.

"Instead of one big aperture from an 11-meter dish, we're going to use several smaller apertures and connect them with digital signal processing," Ingram explained. "A smaller aperture has a wider beam, so the tracking requirement won't be as great. They may pick up interference, especially in tracking a satellite at a low-elevation angle, but because we combine multiple apertures, we can null out the interference."

The arrays individually will not provide the same data rate as NASA's large structures, but having more of them spread out around the world will compensate for that. Network capacity studies show that two ground stations, each with seven 0.75-meter dishes or eight electronically steered antennas, could equal the data capacity of NASA's existing 11-meter dish in Poker Flats, Alaska, at a significantly lower cost.

And because an array does not depend on precisely aiming a dish, each one could potentially communicate with more than one satellite at a time. "What we'd really like to have is a shared antenna resource, in which software is used to separate out the signals," Mandl explained. "As we get more satellites up in space, this will become more important."

In testing performed at Georgia Tech, researchers were able to downlink EO-1 information in the S-band, a frequency used for transmissions at low data rates. They had to develop a special filter to eliminate interference from terrestrial repeater stations of popular satellite radio services.

"We have demonstrated the lower rates in S-band, and, during the upcoming year, we will work on X-band for higher rates," Mandl said. "Ultimately, we would like to demonstrate Ka band, which is in the 27-28 gigahertz range. You could potentially get anywhere from 300 megabits to a gigabit of data in that stream."

To extend satellite reception time, researchers are also examining several technical issues, such as array-based synchronization and optimization of the tilt angles of the planar apertures of the electronically steered antennas. This optimization could quadruple the download capacity for a ground station with eight electronically steered antennas.

If successful, the adaptive array project would give NASA more flexibility in design of future high-data rate satellites that may generate terabits of data on each orbit of the Earth. Reliably downlinking that amount of information will require a new approach, Mandl noted.

Operation and Deployment Experiments Simulator (ODES) engine testing at White Sands Test Facility in New Mexico.

"If you are in the Arctic and the motor moving your dish breaks down, it may take a few weeks to fix it," Mandl said. "If this could be done with no moving parts, using techniques of digital signal processing and software radio, one of the most desirable features will be a high level of reliability. That's important for space applications and locations where you can just put equipment out there and not require an operator or maintenance crew."

Flight Support: Rocket Propulsion Testing

On July 6, 1962, NASA selected the White Sands Test Facility (WSTF) as the site for Johnson Space Center's Propulsion Systems Development Facility. This site was chosen for its isolated location and topography, which minimized the inherent hazards of aerospace propulsion testing to the general population. WSTF began testing rocket engines in 1964. More than 310 engines have been tested, for a total number of firings exceeding 2.1 million.

WSTF's 300 and 400 Propulsion Test Areas were originally constructed to test the engines for the Apollo Command and Service Modules and the Lunar Module. In September 1964, the first firing test of the main rocket engine for the Apollo Command and Service Modules was conducted. The Lunar Module descent engine, which allowed the craft to land softly on the Moon, and the ascent engine, which was used to launch the craft from the lunar surface, were certified for flight after hundreds of firings in the 400 Area. The reaction control system, which consisted of the small thrusters that control the spacecraft attitude, was also certified for flight at WSTF.

Today, six test stands provide vacuum test capability, and three test stands provide ambient testing, 5,000 feet above sea-level, for the Space Shuttle, the ISS, and for other government agency tests.

Stennis Space Center is NASA's primary center for testing and proving flight-worthy rocket propulsion systems for the Space Shuttle and future generations of space vehicles. Having conducted engine testing for 4 decades, Stennis is NASA's program manager for rocket propulsion testing with total responsibility for conducting and managing all NASA propulsion test programs.

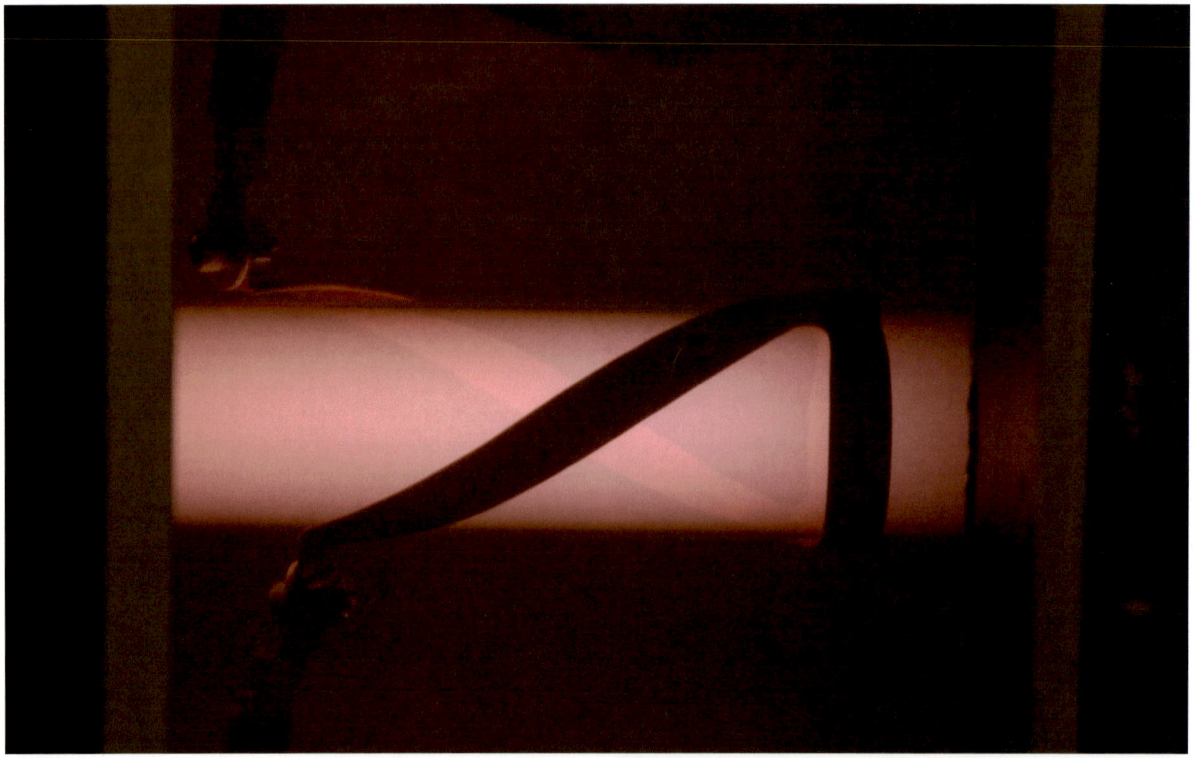

NASA scientists began generating plasma energy in a 9-inch vacuum chamber in NASA's Propulsion Research Laboratory at the Marshall Space Flight Center. In partnership with researchers at the University of Texas at Austin, Johnson Space Center, and the University of Alabama, Marshall scientists are developing innovative magnetic nozzles capable of properly channeling superheated plasma without nozzle deterioration, causing the plasma to accelerate to velocities far faster than those of conventional chemical propulsion systems. Such component technology could support development of next-generation, plasma-propelled spacecraft capable of safely and quickly carrying robotic or human exploration missions deep into the solar system.

Exploration Systems

The Exploration Systems Mission Directorate is responsible for creating new capabilities and supporting technologies that enable sustained and affordable human and robotic exploration. This mission directorate is also responsible for effective utilization of ISS facilities and other platforms for research that support long-duration human exploration.

Plasma Energy Technology to Propel Deep-Space Missions

NASA scientists have begun generating plasma energy in a 9-inch vacuum chamber in NASA's Propulsion Research Laboratory at Marshall. In partnership with researchers at the University of Texas at Austin, Johnson, and the University of Alabama in Huntsville, Marshall scientists are developing innovative magnetic nozzles capable of properly channeling superheated plasma without nozzle deterioration, causing the plasma to reach very high velocities.

Such component technology could support development of next-generation, plasma-propelled spacecraft capable of safely and quickly carrying robotic or human exploration missions deep into the solar system. This could dramatically reduce travel times to Earth's neighboring planets and extend the capabilities of future space exploration missions.

The new research project has two objectives: development of an innovative magnetic nozzle design capable of directing the flow of plasma, and determining how to efficiently eject the plasma from the nozzle to produce the greatest propulsive thrust.

Plasma is a highly conductive medium formed when a gas is heated and ionized—the process in which the gas's neutral atoms shed electrons and acquire a positive charge. When properly channeled through a magnetic nozzle, plasma can be accelerated to velocities dramatically faster than those of conventional chemical propulsion systems.

Propellant in a plasma state can be accelerated with the use of electromagnetic energy sources to increase the propulsion system's specific impulse—the equivalent of a car's gas mileage. Such a nozzle, magnetically insulated against the superheated plasma flow, would enable plasma acceleration at temperatures far beyond those conventional materials can endure.

The second challenge is rooted in the physics of magnetized plasma flow. A plasma propulsion system requires magnetic coils to generate and channel the plasma. These coils produce closed magnetic field lines—circular loops of magnetic energy that form around the power source—and prevent the plasma from detaching and leaving the spacecraft.

The research consortium seeks to test mechanisms that allow the plasma stream—already properly shaped by the magnetic nozzle—to break away from the spacecraft, generating maximum thrust by dispersing the plasma at

exactly the right moment following expulsion from the rear of the spacecraft. Eventually, NASA hopes to adapt this research to develop a new class of rockets incorporating magnetic nozzles and plasma propulsion systems.

NASA Develops Robot With Human Traits

NASA researchers envision futuristic robots that "act" like people, enabling these mechanical helpers to work more efficiently with astronauts. Human-robot cooperation, in turn, will enable exploration of the Moon and Mars, and even large-scale construction in extraterrestrial places. Because human crews will be limited to small teams, astronauts will need robot helpers to do much of each team's work.

Though remotely controlled machines and robots that work entirely on their own are valid goals, a research team at Ames Research Center plans to focus on robots that are partly controlled by people and operate independently the rest of the time.

There are three main areas under development. One is called collaborative control, during which the human being and the robot will speak to one another and work as partners. The second area is building robots with reasoning mechanisms that work similarly to human reasoning. Thirdly, the researchers will conduct field tests of people and robots working together.

Many experiments will occur in a special, indoor laboratory under construction at Ames, featuring a control room with a window looking out on robots working in a large area that will simulate the surface of a moon or planet. The control room will imitate a human habitat on the Moon or Mars.

The robots will help assemble buildings, test equipment, weld structures, and dig with small tools. Human-robot teams will use a checklist and a plan to guide their

Susan Young Lee, lead hardware engineer, and Eric Park, computer scientist, working on a K-10 Rover in one of Ames Research Center's robotics laboratories.

joint efforts. The robot development work will focus on specific tasks essential for basic exploration mission operations including: shelter and work hangar construction, piping assembly and inspection, pressure vessel construction, habitat inspection, resource collection, and transport.

Scientists say human-robot cooperation will result in a better outcome than human- or robot-only teams could accomplish. To make human-machine teaming a reality, a NASA multi-pronged effort is underway to develop robot intelligence. Similar to human thinking, it is designed to improve the mechanical workings of robots and to standardize human-robot communications.

Robots Will Search for Lunar Water Deposits

The Vision for Space Exploration spells out a long-term strategy of returning to the Moon as a step towards sending humans to Mars and beyond. The Moon, so nearby and accessible, is a great place to try out new technologies critical to living on alien worlds before venturing across the solar system.

Whether a Moon base will turn out to be feasible hinges largely on the question of water. Colonists need water to drink. They need water to grow plants. They can also break water apart to make air (oxygen) and rocket fuel (oxygen + hydrogen). Furthermore, water is surprisingly effective at blocking space radiation. Surrounding the base with a few feet of water would help protect explorers from solar flares and cosmic rays. The problem is that water is dense and heavy. Carrying large amounts of it from Earth to the Moon would be expensive. Settling the Moon would be so much easier if water were already there.

Astronomers believe that comets and asteroids hitting the Moon eons ago left water behind. (Scientists believe that Earth may have received its water in the same way.) Water on the Moon does not last long. It evaporates in sunlight and drifts off into space. Only in the shadows of deep, cold craters could an explorer expect to find any, frozen and hidden. Indeed, there may be deposits of ice in such places.

In the 1990s, two spacecraft, Lunar Prospector and Clementine, found tantalizing signs of ice in shadowed craters near the Moon's poles—perhaps as much as a cubic kilometer. The data were not conclusive, though.

To find out if lunar ice is truly there, NASA plans to send a robotic scout. The Lunar Reconnaissance Orbiter, or "LRO" for short, is scheduled to launch in 2008 and to orbit the Moon for a year or more. Carrying six different scientific instruments, LRO will map the lunar environment in greater detail than ever before. LRO's instruments will do many things: they will map and photograph the Moon in detail, sample its radiation environment, and hunt for water.

An eight-legged Scorpion robot prototype test under development at Ames Research Center is just one example of the innovative robotics work being done at that center.

The spacecraft's Lyman-Alpha Mapping Project (LAMP) will attempt to peer into the darkness of permanently shadowed craters at the Moon's poles, looking for signs of ice hiding there. By looking for the dim glow of reflected starlight, LAMP senses a special range of ultraviolet light wavelengths. Not only is starlight relatively bright in this range, but also the hydrogen gas that permeates the universe radiates in this range as well. To LAMP's sensor, space itself is literally aglow in all directions. This ambient lighting may be enough to see what lies in the inky blackness of these craters.

The spacecraft is also equipped with a laser that can shine pulses of light into dark craters. The main purpose of the instrument, called the Lunar Orbiter Laser Altimeter (LOLA), is to produce a highly accurate contour map of the entire Moon. As a bonus, it will also measure the brightness of each laser reflection. If the soil contains ice crystals, as little as 4 percent, the returning pulse would be noticeably brighter.

One of LRO's instruments, Diviner, will map the temperature of the Moon's surface. Scientists can use these measurements to search for places where ice could exist. Even in the permanent shadows of polar craters, temperatures must be very low for ice to resist evaporation. Thus, Diviner will provide a "reality check" for LRO's other ice-sensitive instruments, identifying areas where positive signs of ice would not make any sense, because the temperature is simply too high.

Not far from some permanently shadowed craters are mountainous regions in permanent sunlight, known romantically as "peaks of eternal sunshine." Conceivably, a Moon base could be placed on one of those peaks, providing astronauts with constant solar power—not far from crater valleys below, rich in ice and ready to be mined.

NASA's Desert 'Rats' Test New Gear

Arizona's high desert is not quite as tough on equipment as the Moon or Mars, but few places on Earth can give prototype space suits, rovers, and science gear a better workout.

A NASA-led team headed for sites near Flagstaff, Arizona, in September, to test innovative equipment. Engineers and scientists led the Desert Research and Technology Studies (RATS) team from Johnson and Glenn Research Center. The team included members from NASA centers, universities, and private industry. Their efforts may help America pursue the Vision for Space Exploration to return to the Moon and travel beyond.

The sand, grit, dust, rough terrain, and extreme temperature swings of the desert are attractive, simulating some of the conditions that may be encountered on the Moon or Mars. Crews wearing prototype-advanced space suits used and evaluated the new equipment for 2 weeks.

"For field testing, the desert may be the closest place on Earth to Mars, and it provides valuable hands-on experience," said Joe Kosmo, Johnson's senior project engineer for the experiments. "This work will focus on the human and robotic interaction we'll need for future lunar and planetary exploration, and it will let us evaluate new developments in engineering, science, and operations," he added.

Engineers in the Exploration Planning and Operations Center at Johnson provided mission control-type monitoring of the field tests.

The test equipment included:

- New space suit helmet-mounted speakers and microphones for communications.
- A "field assistant" electric tractor that follows test subjects in space suits, and is guided by space suit-mounted controls.
- A wireless network, for use on other planets, that can relay data and messages among spacewalkers, robots, and rovers as they explore the surface.
- A two-wheeled chariot that is pulled by the electric tractor to carry astronauts.
- "Matilda," an autonomous robotic support vehicle that can retrieve geologic samples.
- Analytical equipment mounted on two mobile geology laboratories.

The Remote Field Demonstration Test Site serves as sort of a dry run of a dry run. Researchers use the rugged terrain and varied climate to test prototype space suits and innovative equipment.

Science

NASA's Science Mission Directorate carries out the scientific exploration of the Earth, Moon, Mars, and beyond; charts the best route of discovery; and reaps the benefits of Earth and space exploration for society. By combining Earth and space science, NASA is best able to establish an understanding of the Earth, other planets, and their evolution, bringing the lessons of our study of Earth to the exploration of the solar system and assuring the discoveries made here will enhance our work there.

Deep Impact Mission

Comets are time capsules that hold clues about the formation and evolution of the solar system. They are composed of ice, gas, and dust, primitive debris from the solar system's distant and coldest regions that formed 4.5 billion years ago. Deep Impact, a NASA Discovery Program mission, is the first to probe beneath the surface of a comet and reveal the secrets of its interior.

At the culmination of the 6-year mission, on July 3, 2005, a 370-kilogram impactor was released from the Deep Impact spacecraft. The spacecraft watched from a safe distance while the impactor collided with comet Tempel 1 at 6.3 miles per second (10 kilometers per second) or 23,000 miles per hour (37,000 kilometers per hour), on July 4. The impact created a magnificent flash of light as an immense cloud of fine powdery material was ejected and subsequently captured in 4,500 images from the spacecraft's cameras.

Scientists continue to analyze the gigabytes of data collected from the 4th of July fireworks in deep space. It is estimated that the crater formed from the impact is between 165 and 820 feet (50 and 250 meters) wide. Analyzed data will be combined with that of other NASA and international comet missions. Results from these missions will lead to a better understanding of both the solar system's formation and implications of comets colliding with planetary surfaces.

Artist Pat Rawlings illustrates the moment of impact and the forming of the crater during the Deep Impact Mission.

Mars Exploration Rover Mission

We're going to overtime—for the third time.

In April 2005, NASA approved up to 18 more months of operations for Spirit and Opportunity, the twin Mars rovers that have already surprised engineers and scientists by continuing active exploration for more than 20 months—well past their 3-month primary mission.

The rovers have proven their value with major discoveries about ancient watery environments on Mars that might have harbored life. Shortly after landing in January 2004, Opportunity found geological evidence of a shallow ancient sea. More than a year later, Spirit found a new class of water-affected rock. The Science Mission Directorate leadership decided to extend the mission through September 2006 to take advantage of having such capable resources still healthy and in excellent position to continue the Mars adventures.

With the rovers already performing well beyond their original design lifetimes, there is a distinct possibility that, at any time, a part could wear out and therefore disable the robotic explorers. Both rovers, however, show

Full frame experiment data record acquired on Sol 494 of Spirit's mission to Gusev Crater, at approximately 16:37:46 Mars local solar time.

no signs of letting up, despite traveling through dust devils and sand traps. Through August 2005, Spirit and Opportunity have explored over 6.5 miles (10.5 kilometers) of Martian terrain.

Cassini-Huygens Mission

The Cassini spacecraft is embarking on a new mission phase that will give it a ringside seat at Saturn—literally.

After concentrating on flybys of Saturn's moons since arriving last year, Cassini began a 5-month study of the stately planet's magnificent rings in April with 12 instruments onboard. Knowing how the rings form and how long they have been there are central questions for the Cassini-Huygens mission.

In a spectacular kickoff to its first season of prime ring viewing, Cassini has confirmed earlier suspicions of an unseen moon hidden in a gap in Saturn's outer "A" ring, known as the Keeler Gap.

The moon, provisionally called S/2005 S1, was first seen in a time-lapse sequence of images taken on May 1, 2005, as Cassini began its climb to higher inclinations in orbit around Saturn. A day later, an even closer view was obtained, which has allowed measurement of its size and brightness.

S/2005 S1 is the second-known moon to exist within Saturn's rings. The other is Pan, which orbits in the Encke Gap of the "A" ring. Imaging scientists had predicted the new moon's presence and its orbital distance from Saturn after a July 2004 sighting of a set of peculiar spiky and wispy features in the Keeler Gap's outer edge. The similarities of the Keeler Gap features to those noted in Saturn's "F" ring and the Encke Gap led imaging scientists to conclude that a small body, a few kilometers across, was lurking in the center of the Keeler Gap, awaiting discovery.

NASA scientists have also concluded that another Saturn moon, Phoebe, is an interloper to the Saturn system from the deep outer solar system.

When Cassini flew by Phoebe on its way to Saturn on June 11, 2004, little was known about the battered, crater-filled moon at that time. During the encounter, scientists got the first detailed look at Phoebe, which allowed them to determine its makeup and mass. As new information unfolded, scientists were able to determine that Phoebe has an outer solar system origin, akin to Pluto and other members of the Kuiper Belt.

"Phoebe was left behind from the solar nebula, the cloud of interstellar gas and dust from which the planets formed," said Dr. Torrence Johnson, a Cassini imaging team member at the Jet Propulsion Laboratory (JPL). "It did not form at Saturn. It was captured by Saturn's gravitational field and has been waiting eons for Cassini to come along."

In this true color view, Mimas, one of the innermost moons of Saturn, drifts along in its orbit, against the azure backdrop of Saturn's northern latitudes.

Phoebe has a density consistent with that of the only Kuiper Belt objects for which densities are known. Phoebe's mass, combined with an accurate volume estimate from images, yields a density of about 100 pounds per cubic foot (1.6 grams per cubic centimeter), much lighter than most rocks, but heavier than pure ice, which is about 58 pounds per cubic foot (0.93 grams per cubic centimeter). This suggests a composition of ice and rock similar to that of Pluto, and Neptune's moon, Triton. Whether the dark material on other moons of Saturn is the same primordial material as on Phoebe remains to be seen.

Meanwhile, new observations have been made about Saturn's largest moon, Titan. Huygens, a European Space Agency probe with six instruments onboard,

Specially calculated Cassini orbits place Earth and Cassini on opposite sides of Saturn's rings, a geometry known as occultation. Cassini conducted the first radio occultation observation of Saturn's rings on May 3, 2005.

landed safely on Titan on January 14, 2005, recording hundreds of megabytes of data during its descent through the atmosphere and while on the surface. Titan is the only known moon in our solar system that has a thick atmosphere. Huygens revealed that the thick atmosphere of this giant moon is rich in organic compounds, whose chemistry may be similar to that of primordial Earth several billion years ago.

"Titan is not just a dot in the sky; these new observations show that Titan is a rich, complex world, much like the Earth in some ways," said Dr. Michael Flasar, the Composite Infrared Spectrometer (CIRS) instrument principal investigator at Goddard. In all, there will be 45 flybys of Titan during the Cassini-Huygens nominal mission, giving scientists more information to unravel the mysteries of its thick atmosphere and other Earth-like processes, such as tectonics, erosion, winds, and perhaps volcanism, which may have shaped Titan's surface.

Swift Mission

Scientists using the Swift satellite—launched on November 20, 2004—and several ground-based telescopes have detected the most distant explosion yet, a gamma-ray burst from the edge of the visible universe.

This powerful burst was detected September 4, 2005. It marks the death of a massive star and the birth of a black hole. It comes from an era soon after stars and galaxies first formed, about 500 million to 1 billion years after the Big Bang. Gamma-ray bursts are the most powerful explosions the universe has seen since the Big Bang. They occur approximately once per day and are brief, but intense, flashes of gamma radiation.

"We designed Swift to look for faint bursts coming from the edge of the universe," said Swift principal investigator, Dr. Neil Gehrels, of Goddard. "Now we've got one, and it's fascinating. For the first time, we can learn about individual stars from near the beginning of time. There are surely many more out there," he added.

The Swift satellite is designed specifically for gamma-ray burst science. Its three instruments work together to observe gamma-ray bursts and afterglows in the gamma-ray, X-ray, and optical wavebands. The Burst Alert Telescope (BAT) monitors the entire sky to catch a gamma-ray burst and calculate an initial position. Within seconds of detecting a burst, Swift will relay the burst's location to ground stations, allowing both ground-based and space-based telescopes around the world the opportunity to observe the burst's afterglow. Armed with the position, the Swift spacecraft autonomously points two other onboard telescopes within their field-of-view, within 90 seconds. All three telescopes watch the gamma-ray burst and afterglow unfold. During Swift's 2-year nominal mission, scientists should have data for approximately 200 gamma-ray bursts to determine their origin and study activities of the early universe.

The Boeing Delta II launch vehicle for NASA's Swift spacecraft is silhouetted against a rosy sky at sunrise, waiting for liftoff.

Detecting Coastal Pollution

Back on Earth, a NASA-funded study of marine pollution in southern California concluded that space-based synthetic aperture radar can be a vital observational tool for assessing and monitoring ocean hazards in urbanized coastal regions.

Image courtesy of Spectrum Astro, Inc.

An artist's rendering of the Swift spacecraft with a gamma-ray burst in the background.

"Clean beaches and coastal waters are integral to southern California's economy and lifestyle," said Dr. Paul DiGiacomo, a JPL oceanographer and lead author of a study recently published in the Marine Pollution Bulletin. "Using southern California as a model system, we've shown existing high-resolution, space-based radar systems can be used to effectively detect and assess marine pollution hazards. This is an invaluable tool for water quality managers to better protect public health and coastal resources," he added.

DiGiacomo and colleagues from JPL; the University of California, Santa Barbara; and the University of Southern California, Los Angeles, examined satellite radar imagery of the state's southern coastal waters. The area is adjacent to 20 million people, nearly 25 percent of the U.S. coastal population.

"The key to evaluating and managing pollution hazards in urban coastal regions is accurate, timely data," DiGiacomo said. "Since such hazards are usually localized, dynamic, and episodic, they're hard to assess using oceanographic field sampling. Space-based imaging radar works day and night, regardless of clouds, detecting pollution deposits on the sea surface. Combined with field surveys and other observations, including shore-based radar data, it greatly improves our ability to detect and monitor such hazards."

The study described three major pollutant sources for southern California: storm water runoff, wastewater discharge, and natural hydrocarbon seepage.

"During late fall to early spring, storms contribute more than 95 percent of the region's annual runoff volume and pollutant load," said JPL co-author Ben Holt. "Californians are accustomed to warnings to stay out of the ocean during and after storms. Even small storms can impact water quality. Radar data can be especially useful for monitoring this episodic seasonal runoff."

DiGiacomo noted that a regional southern California marine water quality-monitoring survey is under way, involving JPL and more than 60 other organizations, including the Southern California Coastal Water Research Project. Its goal is to characterize the distribution and ecological effects of storm water runoff in the region. Space radar and other satellite sensor data are being combined, including NASA's Moderate Resolution Imaging Spectroradiometers (MODIS). The sensors provide frequent observations, subject to clouds, of ocean color that can be used to detect regional storm water runoff and complement the finer resolution, but less frequent, radar imagery.

The second largest source of the area's pollution is wastewater discharge. Publicly owned treatment works discharge daily more than 1 billion gallons of treated wastewater into southern California's coastal waters. Even though it is discharged deep offshore, submerged plumes occasionally reach the surface and can contaminate local shorelines.

Natural hydrocarbon seeps are another local pollution hazard. Underwater seeps in the Santa Barbara Channel and Santa Monica Bay have deposited tar over area beaches. Space-imaging radar can track seepage on the ocean surface, as well as human-caused oil spills, which are often affected by ocean circulation patterns that make other tracking techniques difficult.

Further research is necessary to determine the composition of pollution hazards detected by radar. "From imaging radar, we know where the runoff is, but not necessarily which parts of it are harmful," Holt said. "If connections can be established, imaging radar may be able to help predict the most harmful parts of the runoff."

While the researchers said environmental conditions such as wind and waves can limit the ability of space radar to detect ocean pollution, they stressed the only major limitation of the technique is infrequent coverage. "Toward the goal of a comprehensive coastal ocean observing system, development of future radar missions with more frequent coverage is a high priority," DiGiacomo noted.

Detecting Airborne Pollution

NASA scientists have discovered that pollution could catch an airborne "express train," or wind current, from Asia all the way to the southern Atlantic Ocean.

Scientists believe that, during certain seasons, as much as half of the ozone pollution above the Atlantic Ocean may be speeding down a track of air from the Indian Ocean. As it rolls along, it picks up more smog from air peppered by thunderstorms that bring the pollution up from the Earth's surface.

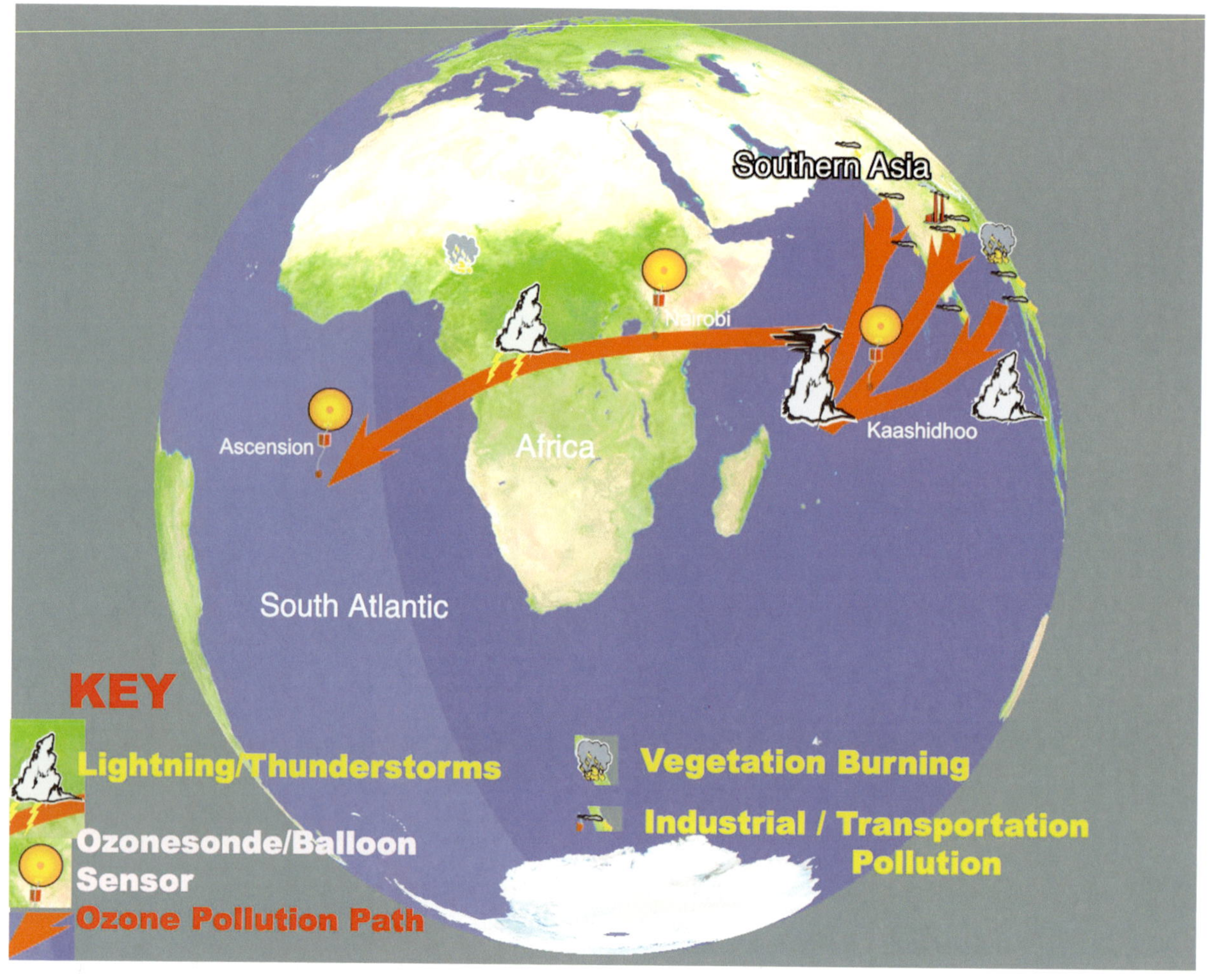

The red arrows on this globe trace the fast track of ozone pollution from Asia as it contributes to the highest ozone episodes found in the South Atlantic.

Bob Chatfield, a scientist at Ames, said, "Man-made pollution from Asia can flow southward, get caught up into clouds, and then move steadily and rapidly westward across Africa and the Atlantic, reaching as far as Brazil."

Chatfield and Anne Thompson, a scientist at Goddard, used data from two satellites and a series of balloon-borne sensors to spot situations when near-surface smog could catch the wind current westward several times annually from January to April.

During those periods of exceptionally high ozone in the South Atlantic, especially during late winter, researchers noticed Indian Ocean pollution follows a similar westward route, wafted by winds in the upper air. They found the pollution eventually piles up in the South Atlantic. "We've always had some difficulty explaining all that ozone," Thompson admitted.

"Seasonal episodes of unusually high ozone levels over the South Atlantic seem to begin with pollution sources thousands of miles away in southern Asia," Chatfield said. Winds are known to transport ozone and pollutants thousands of miles away from their original sources.

Clearly defined, individual layers of ozone in the tropical South Atlantic were traced to lightning sources over nearby continents. In addition to ozone peaks associated with lightning, high levels of ozone pollution came from those spots in the Sahel area of North Africa where vegetation burned. However, even outside these areas, there was extra ozone pollution brought by the Asian "express train."

The scientists pinpointed these areas using the joint NASA-Japan Tropical Rainfall Measuring Mission (TRMM) satellite to see fires and lightning strikes, both of which promote ozone in the lower atmosphere. Researchers also identified large areas of ozone smog moving high over Africa using the Total Ozone Mapping Spectrometer (TOMS) satellite instrument.

They further confirmed the movement of the smog by using sensors on balloons in the Southern Hemisphere Additional Ozonesondes (SHADOZ) network. A computer model helped track the ozone train seen along the way by the SHADOZ balloon and satellite sensors. The scientists recreated the movement of the ozone from the Indian Ocean region to the southern Atlantic Ocean.

Going to 'Extremes'

Hundreds of feet under the Alaskan tundra, Marshall astrobiologist Dr. Richard Hoover ignored the eerie silence of the icy tunnel around him, and even the bones of woolly mammoths and steppe bison jutting from the jagged walls, frozen where they died tens of thousands of years ago.

Forget the fossils.

Hoover was instead poring over pale blue and white patches covering an ice wedge in the tunnel wall. It was a microbial community of bacteria and fungi, growing in total darkness, thriving at temperatures that have hovered below freezing for thousands of years.

For Hoover and his research colleagues, proof of life is the real find, especially in a subterranean tomb, sleeping under ice from the Pleistocene Age. In this unlikely place, they discovered a new life form, a never-before-seen bacterial species they have dubbed *Carnobacterium pleistocenium*. It is roughly 32,000 years old—and it is still alive.

The bacterium—the first fully described, validated species ever found alive in ancient ice—is one of NASA's latest discoveries of an "extremophile." Extremophiles are hardy life forms that exist and flourish in conditions hostile to most known organisms, from the potentially toxic chemical levels of salt-choked lakes and alkaline deserts to the extreme heat of deep-sea volcanoes and hydrothermal vents. NASA and its partner organizations study the

Dr. Elena Pikuta, a scientist at the University of Alabama in Huntsville, and Dr. Richard Hoover, a NASA astrobiologist, lead a team of researchers who recently discovered a new life form: an "extremophile" that lives and thrives in conditions inhospitable to most life on Earth.

potential for life in such extreme zones to help understand the limitations of life on Earth and to prepare robotic probes and, eventually, human explorers to search other worlds for signs of life.

The search for extremophiles is a key element of the Vision for Space Exploration, which aims to reveal unimaginable life forms that could be thriving in conditions few Earth species could tolerate.

"The existence of microorganisms in these harsh environments suggests—but does not promise—that we might one day discover similar life forms in the glaciers or permafrost of Mars, or in the ice crust and oceans of Jupiter's moon, Europa," Hoover noted.

There are approximately 7,000 validly described species of bacteria, though far more are surmised to exist in nature. The vast majority of bacteria are harmless to humans. Only a very few—less than 1 percent of all known species—are dangerous, and many, Hoover noted, are valuable to human life, aiding us in numerous ways: aiding in the production of valuable proteins and life-saving drugs; culturing wine, dairy products, and other foods; and assisting in the biological extraction of gold and other precious metals from ore wastes.

Carnobacterium pleistocenium could offer new breakthroughs in medicine, Hoover said. "The enzymes and proteins it possesses, which give it the ability to spring to life after such long periods of dormancy, might hold the key to long-term cryogenic, or very low-temperature, storage of living cells, tissues, and perhaps even complex life forms," he said.

A collection of NASA's research aircraft on the ramp at the Dryden Flight Research Center in July 1997: X-31, F-15, SR-71, F-106, F-16XL Ship #2, X-38, Radio Controlled Mothership, and X-36.

Aeronautics Research

The Aeronautics Research Mission Directorate is committed to developing tools and technologies that can help to transform how air transportation systems operate, how new aircraft are designed and manufactured, and how our Nation's air transportation system can reach unparalleled levels of safety and security. Such tools and technologies will drive the next wave of innovation, enabling missions to be performed in completely new ways and creating new missions that were never before possible.

NASA has been at the forefront of aeronautics research for decades, and just recently celebrated the 90th anniversary of its predecessor, the National Advisory Committee for Aeronautics (NACA). From March 3, 1915, until its incorporation into NASA on October 1, 1958, NACA provided technical advice to the U.S. aviation industry and conducted cutting-edge research in aeronautics. NACA was created by President Woodrow Wilson, to "direct and conduct research and experimentation in aeronautics, with a view to their practical solution." NASA has continued this tradition.

In the 1920s, NACA engineers developed a low-drag streamlined cowling for aircraft engines, which all aircraft manufacturers then adopted. This innovation resulted in significant operating cost savings. NACA engineers also demonstrated the advantages of mounting engines into the leading edges of multi-engine aircraft wings rather than suspending them, which also became an industry standard.

Through the 1930s, NACA engineers developed several families of airfoils. Many of these were successful as wing and tail sections, propellers, and helicopter rotors used in general aviation and in military aircraft.

During the 1940s, NACA researchers developed the laminar-flow airfoil, which solved the problem of turbulence at the wing trailing edge that limited aircraft performance. The research helped pioneer advances in transonic and supersonic flight. NACA also developed a supersonic wind tunnel, speeding the advent of operational supersonic aircraft and helping to determine the physical laws affecting supersonic flight. In 1945, Robert Jones, one of the premier aeronautical engineers of the 20th century, formulated the swept-back wing concept to reduce shock-wave effects at critical supersonic speeds. Also in the mid-1940s, NACA engineers pioneered research in thermal ice prevention systems for aircraft.

In 1952, NACA's engineers formed the blunt body concept, which suggested that a blunt shape would absorb only a very small fraction of the heat generated during reentry into Earth's atmosphere. The principle was significant for missile nose cones; the Mercury, Gemini, Apollo, and Space Shuttle craft; and unmanned probes. That same year, NACA began studying problems likely to be encountered in space.

In 1954, NACA proposed development of a piloted research vehicle to study the problems of flight in the upper atmosphere and at hypersonic speeds. This led to the development of the rocket-propelled X-15 research airplane.

With NACA's transformation into NASA in 1958, research for space travel became a high-profile endeavor. NASA and Bell Aerosystems Company developed a Lunar Landing Training Vehicle (LLTV) simulator for the Apollo Program. This allowed a pilot to make a vertical landing in a simulated Moon environment. Donald "Deke" Slayton, then NASA's astronaut chief, said there was no other way to simulate a Moon landing except by flying the LLTV.

Four decades of supersonic-combustion ramjet (scramjet) propulsion research culminated in 2004, with two successful flights of the X-43A hypersonic technology demonstrator. The X-43A attained a maximum speed of Mach 9.6, flying freely under its own power. It set world airspeed records for an aircraft powered by an air-breathing engine. The flights proved that scramjet propulsion may be a viable technology for powering future space-access vehicles and hypersonic aircraft.

The Lunar Landing Training Vehicle (LLTV) gave astronauts valuable training in the critical final phases of the descent onto the Moon.

NASA will continue to develop and validate high-value technologies that enable exploration and discovery. The Agency continues its legacy work in aeronautics with breakthrough developments in quieter supersonic and subsonic flight, and autonomous, high-altitude, long-endurance robotic aircraft.

Currently, among its many aeronautics research endeavors, NASA is working toward zero-emission aircraft; smoother, safer airline flights; and elimination of low-visibility-induced accidents.

APEX: Measuring Emissions So That Future Aircraft Fly Cleaner

NASA has been studying various types of emissions from commercial aircraft to develop ways to reduce them and protect the environment. In recent years, fine-particle emissions from aircraft have been identified as possible contributors to global climate changes and to lowering local air quality. These emissions are produced when a hydrocarbon fuel (such as modern jet fuel, which is primarily kerosene) does not burn completely. Incomplete combustion often occurs at the lower power settings used for aircraft descent, idling, and taxiing. This produces fine carbon particles, or soot, as well as particles of non-volatile organic compounds. In addition, engine erosion and small amounts of metal impurities in jet fuel can be emitted in engine exhaust.

Another type of particle emission is formed when exhaust cools, converting volatile aerosols of sulfur compounds and organic compounds to small, solid particles. These types of emissions are not addressed by current international regulations, which focus on visible smoke, but the international community is concerned about the effects that these emissions may have and is identifying possible regulations. In addition, reducing all types of aircraft emissions is necessary for the U.S aircraft industry to remain competitive in the global market.

Recently, Glenn took part in the successful Aircraft Particle Emissions Experiment (APEX). NASA's

The DC-8 airborne laboratory flies three primary types of missions: sensor development, satellite sensor verification, and basic research studies of the Earth's surface and atmosphere.

Data gathered by the DC-8 airborne laboratory at flight altitude and by remote sensing have been used for scientific studies in archaeology, ecology, geography, hydrology, meteorology, oceanography, volcanology, atmospheric chemistry, soil science, and biology.

DC-8 airborne laboratory was used with CFM-56 engines to improve understanding of particle emissions from commercial aircraft engines. It was the first and most extensive set of data obtained about gaseous and particulate emissions from an in-service commercial engine. Many different instruments were used, and a tremendous amount of data was obtained.

NASA scientists ran tests to investigate the effects of thrust and fuel type. The team used different engine operating settings to vary thrust, and three different fuels were used: a typical jet fuel, a fuel with high sulfur content, and a fuel with high aromatic compound content. In addition, the Environmental Protection Agency ran tests to simulate landing-takeoff cycles to study the emissions that would be created at an airport. It was the first time that so many different groups had worked together to study so many different aspects of the emissions from commercial aircraft engines.

Smoothing Out the Skies

Passengers on a Delta Air Lines jet could have a smoother ride, thanks to NASA-developed technology. Delta is installing a special production-prototype radar, which can detect turbulence associated with thunderstorms, on one of its B737-800 aircraft. The radar, called the Turbulence Prediction and Warning System (TPAWS), was developed for NASA's Aviation Safety and Security Program at Langley Research Center.

NASA teamed with Delta Air Lines, of Atlanta; AeroTech Research (USA), Inc., of Newport News, Virginia;

and Rockwell Collins, of Cedar Rapids, Iowa, for the in-service evaluation of the radar unit, which also includes turbulence hazard prediction capabilities.

"The TPAWS technology is an enhanced turbulence detection radar system, which detects atmospheric turbulence by measuring the motions of the moisture in the air," said Jim Watson, the TPAWS project manager. "It is a software signal processing upgrade to existing predictive Doppler wind shear systems, also developed by NASA, that are already on airplanes."

The idea behind the turbulence detection system is to give flight crews advanced warning, so they can avoid turbulence encounters or advise flight attendants and passengers to sit down and buckle up to avoid injury. Turbulence encounters are hazardous, and they cost the airlines money and time in the form of re-routing flights, late arrivals, and additional inspections and maintenance to aircraft. Atmospheric turbulence encounters are the leading cause of injuries to passengers and flight crews in non-fatal airline accidents. Federal Aviation Administration statistics show an average of 58 airline passengers are hurt in U.S. turbulence incidents each year. Ninety-eight percent of those injuries happen because people do not have their seatbelts fastened.

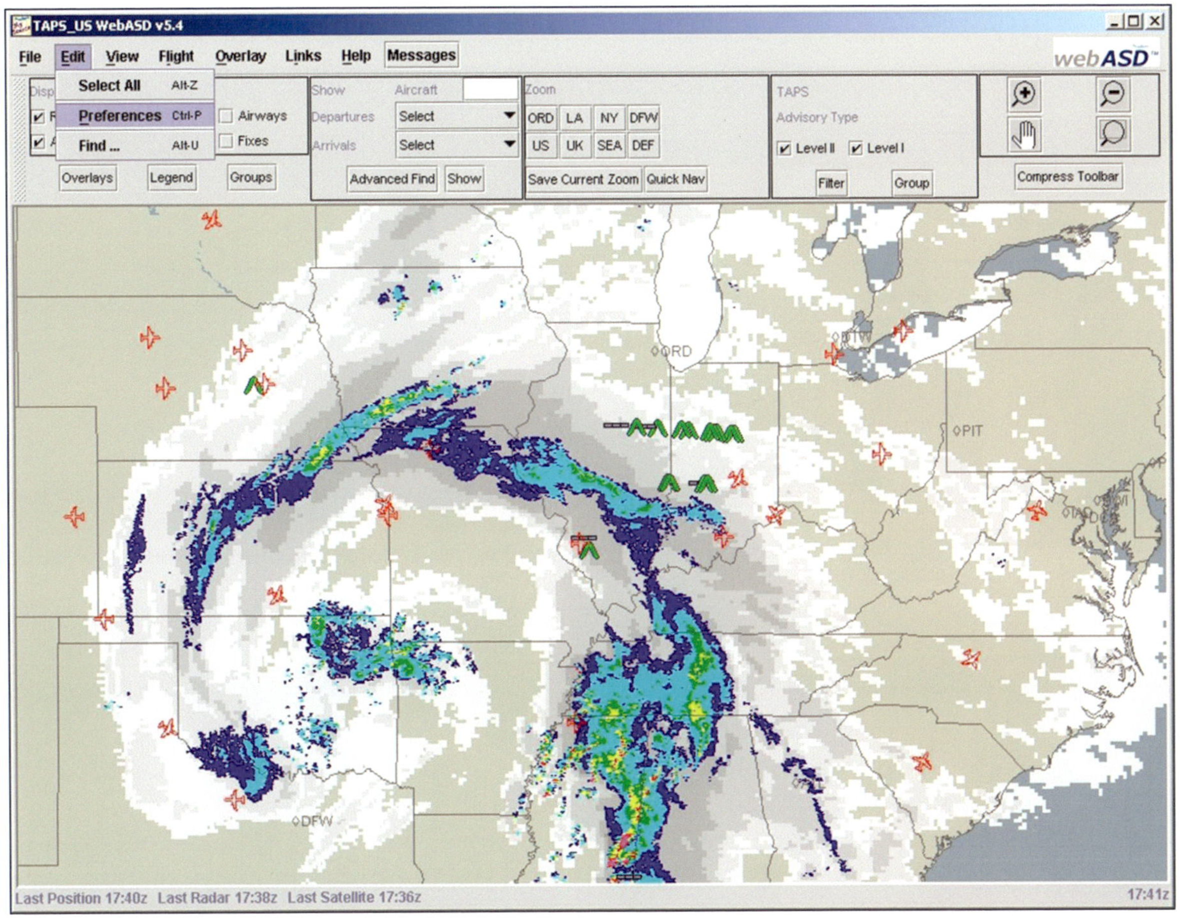

Image courtesy of AeroTech Research (USA), Inc.

Dispatcher's display of turbulence-encounter reports integrated with weather data. These reports are used to safely guide Delta Air Lines flights in real time.

NASA researchers say the TPAWS radar can detect about 80 percent of all atmospheric turbulence encounters. It can also detect thunderstorm-related turbulence at an average of 3 to 5 minutes ahead of the aircraft. According to studies done by Dryden Flight Research Center engineers, it takes a little more than a minute and a half to get 95 percent of passengers seated, carts stored, and flight attendants secured. Delta flight crews will use and evaluate the technology during regularly scheduled flights in the United States and South America. The prototype is expected to fly for 6 to 9 months.

Researchers from NASA, the companies involved, and the Federal Aviation Administration, will evaluate interim and final results of the turbulence prediction radar system. If the evaluation is successful, the technology may be adopted for new and existing aircraft.

NASA has already tested TPAWS on a research aircraft based at Langley. The TPAWS-equipped plane searched for turbulence activity around thunderstorms for 8 weeks. The jet flew within a safe distance of storms, so researchers could experience the turbulence and compare the radar prediction to how the plane responded to the encounters. After one severe patch of turbulence, a NASA research pilot said his confidence in the enhanced radar had "gone up dramatically," since the plane's weather radar had shown nothing at the same time the TPAWS display had shown rough skies ahead. ❖

EDUCATION NEWS AT NASA

NASA's unique missions to space are the result of scientific expertise and technical excellence, qualities that are dependent on sound educational backgrounds. NASA gives back to the educational community in order to groom the next generation of explorers. The following pages show some of the partnerships between the Space Agency and various educational communities.

Education News at NASA

The main goals of NASA's Education Program are to inspire and motivate students to pursue careers in science, technology, engineering, and mathematics, and to engage the public in shaping and sharing the experience of exploration and discovery. These goals are carried out by supporting education in national and international schools, as well as public outreach efforts.

NASA's commitment to education places special emphasis on these goals by increasing elementary and secondary education participation in NASA programs; enhancing higher education capability in science, technology, engineering, and mathematics disciplines; increasing participation by underrepresented and underserved communities; expanding e-Education; and expanding participation with the informal education community.

The Office of Education will continue to support NASA's strong historical role in education at all levels, with linkages to NASA research as a central part of its focus.

The "Circle of Life" camp, run by the National Federation of the Blind, took place, in part, at the Goddard Space Flight Center.

Opening Blind Eyes to Science

Steven and Amelia are blind, yet both were able to read the temperature with a thermometer and measure precipitation with a rain gauge at a science camp last summer. Their secret?

Steven and Amelia, along with 10 other blind students, ages 11 to 14, were using a talking thermometer, a Braille-marked rain gauge, and other tools identified by NASA for use by the visually impaired.

The tools got their first major test in July at the end of the week-long "Circle of Life" camp hosted by the National Federation of the Blind and sponsored by NASA. On the camp's final day, students visited a pond and forest area situated at Goddard Space Flight Center, where they made observations of the soil, vegetation, weather, and birds.

For many of the kids, it was the first time they had used observation instruments specifically geared toward the senses of sound and touch, rather than sight.

"I didn't know they existed," said Steven, now a ninth-grader at a science-oriented high school in New York. "It was amazing to 'see' the technology."

Goddard soil scientist Elissa Levine has been leading NASA's effort to introduce various blind-friendly gadgets to the visually impaired community. Her work is aimed at making a variety of activities more accessible to the blind, including GLOBE, a NASA-sponsored science education program in which K-12 students around the world take measurements of soil, land cover, air, water, and living things.

"I have been working with the GLOBE program for many years and am aware of how effective it is as a learning experience, which made me interested in seeing if it would be as effective for blind students as well," Levine said.

The new instruments include two kinds of talking thermometers, one for the air and a meat thermometer that can be inserted into the soil. There is also a talking compass and a talking sensor that analyzes soil color. A graduated cylinder with a floating plug serves as a rain gauge. The plug moves up or down, depending on the volume of water, and is attached to a plastic measuring scale marked in Braille.

Unlike other sciences that are sometimes more abstract, Earth science provides plenty of convenient opportunities for interactive, nonvisual activities, such as listening to birds or rubbing soil between one's fingers.

"The best thing about Earth science is that there is a natural laboratory right outside the classroom door," Levine noted.

Promoting interest in science—Earth or otherwise—among the more than 93,000 estimated blind school-age children in the United States is as much about educating teachers as it is kids, according to the "Circle of Life" camp's lead instructor, Robin House, who says teachers often underestimate the potential of blind students.

"Many times blind students are left out of sciences and math because some educators think, 'Oh, this is too difficult, they couldn't possibly grasp these concepts,'" said House, who herself is blind. "The idea of this particular camp was a little bit of exposure in all the areas of science to get kids going, 'I can do science, I can do it. I can become a scientist if I want to.'"

In the days preceding their visit to Goddard, the campers dissected a dogfish shark and dug for dinosaur fossils at the Maryland Science Center, explored seashells with blind shell expert Geerat Vermeij, took a boat ride on the Chesapeake Bay, and listened to sounds from space with blind physicist Kent Cullers.

For 12-year-old Amelia, the camp was a rare chance to experience science up close and personal.

"There are a lot of tools out there that blind people can use to investigate science. Blind people can pretty much do anything, but they just do it differently."

Steven has a simple, straightforward message for both students and teachers: "Blindness doesn't have to be a barrier. Being blind doesn't stop you from having a brain and doing science."

NASA works to capture the educational potential of its robotics missions by supporting educational competitions and events, facilitating robotics curriculum enhancements at all educational levels, and maintaining a Web site clearinghouse of robotics education information.

The camp was the first step in the National Federation of the Blind's initiative to create a National Center for Blind Youth in Science. Another camp held last summer called "Rocket On!" challenged blind high school students to develop, build, and launch a 12-foot rocket from NASA's Wallops Flight Facility.

NASA Sponsors Eighth Annual Botball Robotics Competition

Thirty-three Northern California middle and high school teams demonstrated their robotics skills at the eighth annual northern California Botball Robotics Tournament in April at Santa Clara University.

Botball is a robotics program designed to engage students in learning science, technology, engineering, and

Students participating in the NASA Explorer School Program join NASA's mission of discovery through educational activities and special learning opportunities tailored to promote careers in science, mathematics, and engineering.

math. Students are given 7 weeks to design, build, and program two microcontrolled robots with LEGO structures, to compete in a fast-paced regional tournament consisting of head-to-head double elimination rounds. By building robots, students are exposed to high-tech equipment, gain knowledge of project design and computer programming, and develop team problem-solving skills. Unlike many robotics competitions, though, Botball robots are programmed in the "C" programming language and, therefore, use no remote controls. Game play is based solely on the skill of the team programmers. Students also can compete in the creation of Web sites to document their team progress.

The tournament is presented by the KISS Institute for Practical Robotics, based in Norman, Oklahoma, in conjunction with Santa Clara University and NASA's Robotics Education Project at Ames Research Center. The NASA Robotics Education Project is supported through NASA's Science Mission Directorate and is directed by David Lavery, program executive for Solar System Exploration.

NASA Explorer School Concept Goes International

A NASA education initiative designed to bring science, technology, engineering, and mathematics learning and activities to U.S. educators, students, and families is going abroad to the Netherlands.

NASA Chief Education Officer, Dr. Adena Williams Loston, signed an agreement among NASA, the European Space Agency (ESA), and the Dutch Ministry of Education, Culture, and Science, establishing the Delta Researchers Schools (DRS) Program.

Patterned after the NASA Explorer School (NES) Program, the DRS Program will identify and develop innovative methods to inspire Dutch primary school students to pursue careers in mathematics and science. The program will focus on stimulating the interest of children between the ages of 9 and 12. It also will generate positive awareness of human space flight, the International Space Station, and other international cooperative projects. The Dutch Ministry of Education, Culture, and Science initially will launch, manage, and fund activities for selected schools for 3-year periods.

The program's name derives from the ESA Delta Mission, which was conducted as part of the International Space Station/Soyuz crew-exchange missions flown in April 2004. As part of the cooperation, NASA will provide opportunities for Dutch teachers to participate in summer NES workshops at NASA centers. Teachers will acquire new resources and technology tools using NASA's unique content. NES officials will be available for content consultation and to coordinate distance-learning capabilities to support the DRS Program. One ham-radio opportunity will be scheduled each year for Dutch students to talk with the Earth-orbiting Space Station crew.

The NES Program has provided more than 70,000 U.S. elementary, middle, and high school students with information and interactive activities on future careers, to fulfill the Vision for Space Exploration.

NASA Summer High School Program Celebrates Silver Anniversary

The NASA Summer High School Apprenticeship Research Program (SHARP) recently celebrated 25 years of selecting high-achieving students to serve as apprentices in a variety of NASA professions. Students, representing nearly every state, Puerto Rico, and American Samoa have participated in the education program.

Created in 1980, SHARP is designed to attract and increase participation among underrepresented students in science, technology, engineering, and mathematics. The apprenticeship runs approximately 8 weeks during the summer and includes an hourly stipend for students. It engages them in research opportunities and reinforces educational excellence. The program also seeks to enrich and inspire the students by promoting interaction within their academic, workplace, and social environments.

All 10 NASA centers and several universities participate in SHARP, including the California State University, Los Angeles; the Georgia Institute of Technology; North Carolina Agricultural and Technical State University; Hampton University, Hampton, Virginia; the University of Michigan, Ann Arbor; the University of New Mexico, Albuquerque; and the University of Wisconsin-Madison.

There is a commuter component for students residing within a 50-mile radius of a NASA center and a residential for those living on a participating university campus.

Hands-on research gives college students work experience in the NASA environment, and in the field.

NASA and University of North Dakota Sign DC-8 Agreement

NASA has signed a cooperative agreement with the University of North Dakota (UND), Grand Forks, to house and operate the Agency's DC-8 jet aircraft. The purpose of the agreement is to create a National Suborbital Education and Research Center (NSERC) at the university with the DC-8 suborbital laboratory as the centerpiece. The agreement is intended to expand the science conducted using the DC-8 and enhance hands-on educational opportunities for students.

Transfer of the aircraft to the university is targeted for fall 2005, pending completion of a safety review. The aircraft will be housed at the Grand Forks Air Force Base.

The DC-8 has been part of NASA science programs since 1986. It has supported satellite validation, Earth science studies, and the development of remote-sensing techniques for space-based observing systems. It has operated from several NASA centers and has been deployed worldwide to support research including ozone depletion, tropical rainforest ecology, hurricane studies, and ice sheets. Its most recent campaign was to New England last January to support arctic ozone studies and validation of NASA's Aura satellite.

UND is home to the Northern Great Plains Center for People and the Environment, which will have oversight for the NSERC. The university is also home to the largest collegiate aviation program in the United States. It maintains and operates 120 aircraft throughout the country, including 80 aircraft in Grand Forks. The aircraft support aviation, atmospheric sciences, space studies, and computer science education activities.

Through the agreement, the university will maintain, operate, and manage educational and science flight missions. NASA retains operational control including safety, airworthiness, and mission management. ❖

PARTNERSHIP SUCCESSES

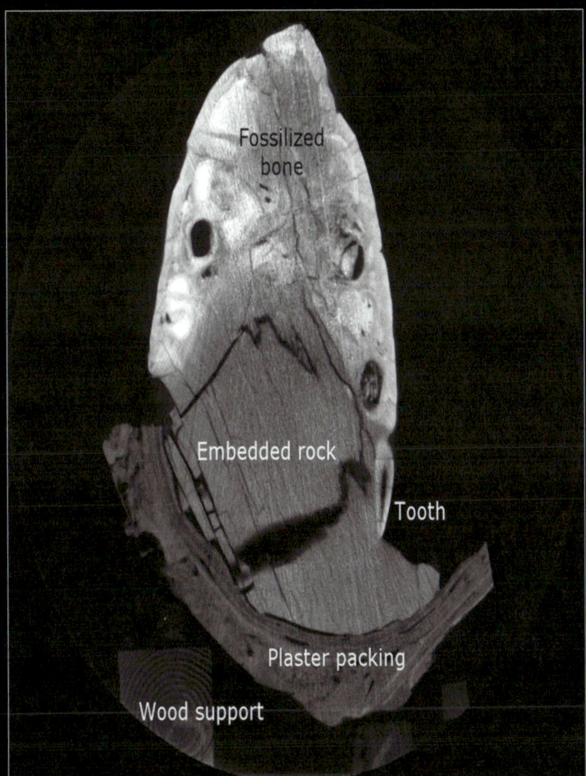

NASA cultivates relationships with private industry, academia, and other government agencies to bring its science back down to Earth. By contributing time, facilities, and a wealth of technical expertise, NASA enriches the lives of people everywhere—though not all of these partnerships result in a commercial spinoff. The following pages show some of the many benefits of these partnerships.

Partnership Successes

New Uses for Ultrasound Technology Aboard the ISS and on Earth

Astronauts onboard the International Space Station (ISS) are using ultrasound to look inside themselves as part of a NASA project called ADUM, short for "ADvanced Ultrasound in Microgravity." Dr. Scott Dulchavsky, a surgeon at the Henry Ford Hospital in Detroit, heads the project. His team, which includes co-investigators Doug Hamilton, Shannon Melton, and Ashot Sargsyan, of Wyle Laboratories, Inc., in Houston, is studying how ultrasound can be used to diagnose medical problems onboard spacecraft.

Here on Earth, doctors can look at broken bones with an X-ray machine; they can look for tumors with a CAT scanner; and they can examine the brain with an MRI. None of those bulky instruments is available on any NASA spacecraft. There is, however, an ultrasound machine onboard the ISS. Ultrasound offers several advantages. Compared to other diagnostic imaging tools, ultrasound machines are compact and lightweight. This is important on cramped spacecraft where every ounce of payload costs money to launch. Furthermore, ultrasound images appear instantly. Crews need not wait for films to be developed. Ultrasound can identify problems quickly.

Typically, ultrasound has been used to look at internal organs. It is often used to examine fetuses. But Dulchavsky and his team have been expanding its repertoire. They are working out ways to look at eyes, teeth, lungs, bones, and muscles. They believe that ultrasound can be used for about two-thirds of a list of approximately 500 medical conditions that might hypothetically occur on a spacecraft.

In some cases, ultrasound works even better in space than it does on Earth, because in low gravity, internal organs move around. The result is that organs often end up closer to each other. Sound waves move from one to the other with less distortion, providing a clearer ultrasound picture.

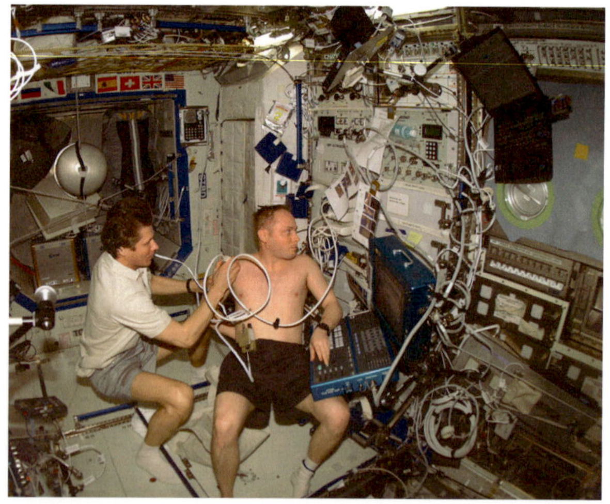

Onboard the International Space Station, Gennady Palalka performs an ultrasound examination on Mike Fincke.

Traditionally, ultrasound probes are operated by technicians with several hundred hours of training. Astronauts only receive about four hours of training. As the astronauts work the probe, they are in constant contact with experts on the ground.

This technique, non-doctors using ultrasound to obtain diagnostic quality pictures under the guidance of remote experts, turns out to have important applications on Earth—on battlefields, for instance, or in rural areas where doctors are far away.

The process has already been used successfully on the ground, in the locker room of the Red Wings, Detroit's hockey team. "Players get hurt a lot in NHL games," says Dulchavsky, a fan. "Last season, we trained one of their trainers to use the probe. It worked famously."

It is also being used by Major League Baseball's, Detroit Tigers, as well as by the Olympic Committee for assessing injuries in snowboarding accidents and ski jump mishaps.

Dulchavsky and his colleagues are analyzing their data. The next step, he says, is to put together a program that will teach the astronauts to do more and more on their own. This would enable ultrasound to be used even on long-range exploration missions, like trips to Mars, where guidance from the ground is less practical. The ADUM project is significant, says Dulchavsky, because it has pushed the limits of what ultrasound technology can do. He and his colleagues plan to push those boundaries even more.

Atomic Oxygen Restores Artwork

NASA research into the damage to satellites caused by atomic oxygen in low-Earth orbit has led to a new way to restore damaged artwork. Atomic oxygen is an elemental form of oxygen that does not exist in Earth's atmosphere. In space, however, it is common in the area where satellites orbit Earth. There, it exposes satellites and spacecraft to damaging corrosion. Researchers at Glenn Research Center study these damaging effects in order to find materials and methods to extend the lifetime of communication satellites, the Space Shuttles, and the International Space Station.

While developing methods to prevent damage from atomic oxygen, researchers discovered that atomic oxygen could remove layers of soot or other organic (carbon-based) materials from a surface. Because atomic oxygen will not react with inorganic oxides, such as most paint pigments, it could be used to restore paintings damaged by soot. For paintings containing organic pigments (which could be damaged by the atomic oxygen), the exposure could be carefully timed so that the removal would stop just short of the paint pigment.

It has been estimated that, worldwide, an average of one collection or gallery suffers fire damage every day, and paintings damaged by charring are very resistant to traditional cleaning techniques. Current processes used to restore artwork generally use chemical solvents to remove dirt, varnish, and thin layers of soot. With damage from heavy deposits of soot, or even charring or graffiti, these techniques are not effective.

The left photograph was taken after the Cleveland Museum of Art staff used acetone and methylene chloride to clean and restore the painting. The right photo was taken after Glenn researchers used the atomic oxygen technique to clean the painting.

In 1996, Glenn researchers Bruce Banks and Sharon Miller were contacted by conservators from the Cleveland Museum of Art about the possibility of using new restoration techniques at the NASA Electro-Physics Branch. The Electro-Physics Branch had facilities (simulating the low-Earth orbit environment) that produce atomic oxygen, which could potentially be used to restore artwork. The first tests were done on two religious paintings damaged by an arson fire at St. Alban's Church in Cleveland Heights, Ohio. Both paintings were found to be unsalvageable by conventional art restoration wet chemistry techniques and were provided to NASA to test its atomic oxygen cleaning process. The technique not only removed the soot, but it cleaned the paintings so effectively that colors that had been faded by time were brighter, and more detail could be seen than before the fire. The success encouraged the art community to risk more important works of art to test the technique further.

In 1998, the atomic oxygen restoration system had its first big success. The Andy Warhol painting "Bathtub," estimated to be worth several hundred thousand dollars, had been kissed by a vandal during a party at the museum. Because Warhol had not varnished the painting, conservators at the Carnegie Museum of Art in Pittsburgh were concerned that traditional solvents would cause the lipstick to soak into the painting and make things worse.

Until they heard about Glenn's atomic oxygen restoration system, conservators had been resigned to keeping the painting in permanent storage. Glenn researchers built a portable version of the atomic oxygen device and transported it to the museum. Preliminary tests were done outside of the viewing area; then the device was used successfully to remove the lipstick smudge.

In addition to the St. Alban and Andy Warhol paintings, a fire-damaged Roy Lichtenstein ink drawing on paper has been cleaned along with two smoke-damaged paintings from St. Stanislaus Church in Cleveland.

In the upper reaches of the atmosphere, about 200 to 500 miles above the Earth, atomic oxygen is created by exposure to intense solar ultraviolet light. Oxygen molecules (two oxygen atoms bonded together as O_2) decompose into two separate oxygen atoms, or atomic oxygen. Because the unpaired atoms react very easily with other materials, they are very destructive to spacecraft and satellites, but very beneficial for cleaning Earthly surfaces. Atomic oxygen can remove any organic coating (a compound containing carbon) from a painting that contains inorganic paint pigments by reacting with the organic coating. This forms a gaseous byproduct while leaving the inorganic pigments undisturbed.

The process is environmentally "green." No solvents are used or produced and the only byproducts of the atomic oxygen formation and reaction processes are trace amounts of ozone, carbon monoxide, and carbon dioxide.

Atomic oxygen treatment can restore works of art when there is surface char or when there are defacements or contaminants on surfaces on which solvents cannot be used. The invention can remove all types of organic protective coatings uniformly over the surface without physical contact, which could alter the painting. Low spots and high spots on the painting surface can be cleaned equally well.

Art restorers are extremely cautious, because careless cleaning could easily damage the unique and highly valuable paintings and prints that they are called on to restore. So, in recent years, the process has been tested to determine its ability to safely treat the range of media typically used by artists (oil paint, acrylic paint, acrylic gesso, watercolors, pen and ink, and others). In 2001, validation testing was completed, and the process was deemed to be acceptable for functional art restoration and ready for licensing. Atomic oxygen is now safe for cleaning artwork where conventional techniques have not been effective.

The experience gained from studying atomic oxygen damage to spacecraft led to its use for restoring works of art. Now NASA is applying the experience gained restoring art to further understand space exposure, as well as to a variety of other medical and industrial applications. One interesting application is the removal of biocontaminants from the surfaces of orthopedic implants (such as artificial hip joints) prior to surgery. NASA researchers continue to turn the damaging effects of atomic oxygen on spacecraft into beneficial uses here on Earth.

Human Genome Activity

Results of NASA scientists' recent research on human DNA are enhancing our knowledge about human genetics and may help us to better understand human diseases.

Scientists at Ames Research Center, in collaboration with scientists from Yale University, have designed a complete map of all gene activities in human tissue.

"As a result of this research, we have a more comprehensive view of human gene activity. This will enable scientists to better understand gene responses to space flight and help NASA ensure astronauts' well-being during long-duration space flights or exploring the Moon and Mars," said Dr. Viktor Stolc, director of the Genome Research Facility at Ames.

Using advanced technology, researchers attached short pieces of DNA that recognized sequences in the human genetic blueprint, called the genome, to specially patterned glass slides. These slide arrays were used to measure levels of ribonucleic acid, biochemical copies of the DNA produced when genes are activated to make proteins. Researchers used high-resolution imaging technology to look at human genome to see previously unknown and unmapped activities.

"In our previous work, we mapped the genome of a fruit fly, which is a model organism for biological processes," Stolc said. "Now, we are making an essential step towards understanding human illness by mapping out the complete human genome activity. We discovered many DNA sequences, originally counted as non-functioning segments, actually do encode active genes. These findings are going to allow us to dissect human diseases and help us find new treatments," he added.

Based on a pilot experiment that studied the genome of a fruit fly, the method used by Stolc and Michael Snyder of Yale proved successful, even on human DNA sequences that are much longer and more complex. "We had to overcome bioinformatics challenges, but at the end, we were rewarded with a comprehensive picture of human tissue DNA," Stolc noted. In a separate-but-related research effort, Stolc and his team of genome researchers collaborated with the University of California, San Francisco (UCSF), to create the first complete map of the gene activity of flagella (microscopic hairs) on single-cell microorganisms—a partnership that aims to provide more insight into a life-threatening kidney disease that has affected more than 600,000 Americans and an estimated 12.5 million people worldwide.

"Hairs on the cell surface in algae are virtually identical to human cilia, short hairs inside a human body that drive fluids across the surface of a cell. Understanding the genetic make-up of the cilia-like structures, through studying the complete genetic code in microorganisms, helps scientists gain a better understanding of polycystic kidney disease (PKD)," said Stolc. According to the PKD Foundation, of Kansas City, Missouri, PKD is the most common genetic, life-threatening disease.

"Gene PKGD1 encodes a protein that is responsible for PKD among human subjects and is also a component of primary cilia in the kidney," Stolc said.

"In fact, several of the genes identified in the study are known to be involved in control of cell behavior and other tissues, raising the possibility these same genes could be the missing link between cilia and polycystic kidney symptoms," said Dr. Wallace Marshall, assistant professor of biochemistry and biophysics at UCSF.

Scientists believe understanding cilia functions may lead to the development of countermeasures to prevent PKD, which is one cause of kidney stone formation.

The study results are published in the March 8, 2005 online version of the Proceedings of the National Academy of Sciences journal.

Tsunami Recovery and Relief Efforts

Imagery from three NASA spaceborne instruments has shed valuable insight into the Indian Ocean tsunami that resulted from the magnitude-9 earthquake southwest of Sumatra on December 26, 2004.

The images offer several unique views of portions of the affected region. The data are and will be used by scientists and government agencies to assist with disaster recovery, mitigate the effects of future natural hazards, and increase our understanding of how and why tsunamis strike. The data were acquired by the Multi-angle Imaging SpectroRadiometer and the Advanced Spaceborne Thermal Emission and Reflection Radiometer instruments on NASA's Terra spacecraft, as well as from the Shuttle Radar Topography Mission.

The Multi-angle Imaging SpectroRadiometer imagery includes the only known animations produced by a remote-sensing instrument to capture tsunami waves in motion as they make landfall. The image set and animations were collected December 26 as Terra passed over the eastern Indian coast about an hour and a half after the first waves hit shore. The first animation shows tsunami waves breaking along the shores of the Indian state

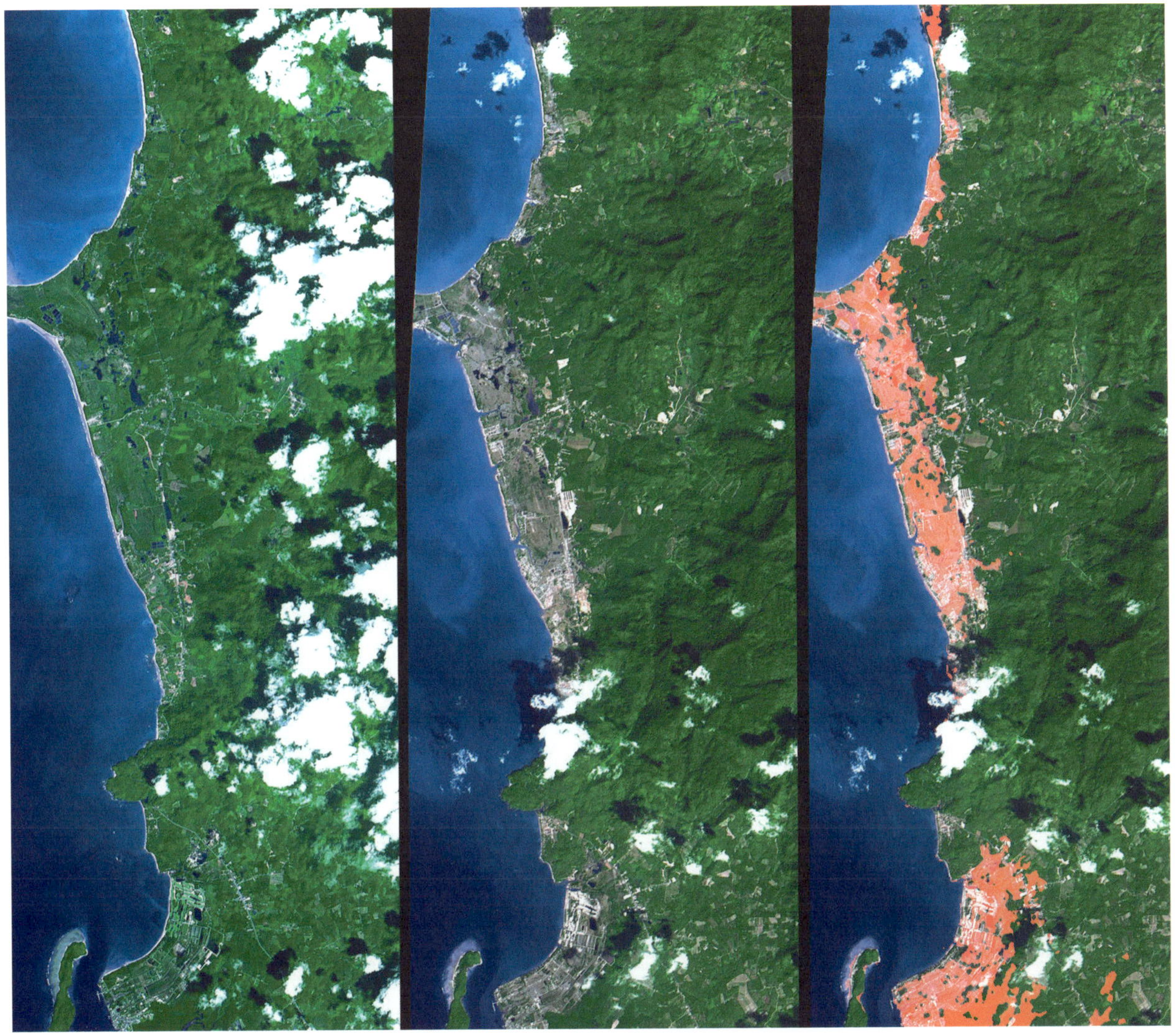

The Indian Ocean coastline near Phuket, Thailand, is a major tourist destination that was in the path of the tsunami produced by a giant offshore earthquake on December 26, 2004. These simulated natural-color Advanced Spaceborne Thermal Emission and Reflection Radiometer images show a 17-mile-long stretch of coastline 50 miles north of the Phuket airport on December 31 (middle) and also 2 years earlier (left). The changes along the coast show (changing from green to grey) where the vegetation was stripped away by the tsunami. The image on the right shows areas, in red, that have elevations within 33 feet of sea level. This elevation information was supplied by the Shuttle Radar Topography Mission. The red areas appear to include most of the tsunami-inundated areas.

of Andhra Pradesh, near the mouth of the Godavari River. Because the instrument's multiple cameras imaged the coast over several minutes and the waves were unusually large, the instrument captured unique time-lapse imagery of them. The still images show frames from four of the instrument's cameras, and span a period of about 2.5 minutes. In the second animation, waves break further to the south, near the mouth of the Krishna River.

The data indicate the location and timing of some of the waves; their angle, relative to the shoreline; and their speed, estimated from these data to be about 48 miles per hour. Together, with measurements of ocean depth, these data can be used to refine models of how tsunamis originate and travel. Better understanding of how tsunamis interact with coastal areas is one factor needed to improve near-real-time forecasts of tsunami arrival times and effects, and to reduce damage from such waves in the future.

Terra's Advanced Spaceborne Thermal Emission and Reflection Radiometer instrument acquired images of the area around Phuket on the Indian Ocean coast of Thailand, on December 31. A pair of simulated natural-color images shows a 17-mile stretch of coastline north of the Phuket airport on December 31, 2004, compared to an image acquired 2 years earlier. The changes along the coast are self-evident, clearly indicating the extent of vegetation stripped by the waves.

The images are being used to create damage assessment maps for the U.S. Agency for International Development, Office of U.S. Foreign Disaster Assistance.

The separate image trio depicts these same before and after views and contrasts them with a third view created with Shuttle Radar Topography Mission data. Elevations below 10 meters are highlighted in red, and include most of the areas inundated by the tsunami, through offshore ocean depth variations, coastal landforms, distance from the coast, and additional factors. Elevation measurements provided by the Shuttle Radar Topography Mission give a general indication of areas at risk and can help planners better predict which areas of a region are in the most danger and help develop mitigation plans in the event of particular flood events.

The Shuttle Radar Topography Mission has also produced a color-coded, shaded relief map of the island nation of Sri Lanka, highlighting regions below 10 meters in elevation. The data were collected during the 11-day Space Shuttle mission in February 2000 and released publicly in July 2003. The low-coastal elevations extend 3.1 to 6.2 miles inland and are especially vulnerable to flooding associated with storm surges, rising sea level, and tsunami conditions.

The Multi-angle Imaging SpectroRadiometer was built and is managed by the Jet Propulsion Laboratory (JPL); Japan's Ministry of Economy, Trade, and Industry built the Advanced Spaceborne Thermal Emission and Reflection Radiometer; and Goddard Space Flight Center manages the Terra satellite. The Shuttle Radar Topography Mission is a collaboration of NASA, the National Geospatial-Intelligence Agency, and the German and Italian space agencies.

NASA also worked to provide Moderate Resolution Imaging Spectroradiometer (MODIS) and Earth Observing-1 (EO-1) satellite Hyperion sensor data collections and historical data to the U.S. Navy, to help them safely bring ships involved in the relief effort into ports damaged by debris and sediments.

Forecasting Earthquakes

Shifting from tsunamis to earthquakes, a NASA-funded earthquake prediction program has accurately predicted the locations of 15 of California's 16 largest earthquakes this decade, including the magnitude-6 quake that shook the state's Parkfield region in late 2004.

The 10-year Rundle-Tiampo Forecast was developed by researchers at the University of Colorado (now at the University of California, Davis) and JPL, with funding from NASA and the U.S. Department of Energy.

"We're elated our computer-modeling technique has revealed a relationship between past and future earthquake locations," said Dr. John Rundle, director of the Computational Science and Engineering initiative at the University of California, Davis. "We're nearly batting a thousand, and that's a powerful validation of the promise this forecasting technique holds," he added.

Of 16 earthquakes, magnitude 5 and higher since January 1, 2000, 15 fall on "hotspots" identified by the forecasting approach. The forecast "scorecard" uses records of earthquakes from 1932 onward to predict locations most likely to have quakes of magnitude 5 or greater between 2000 and 2010. According to Rundle, small earthquakes of magnitude 3 and above may indicate stress is building up along a fault. While activity continues on most faults, some of those faults will show increasing numbers of small quakes, building up to a big quake, while some faults will appear to shut down. Both effects may herald the possible occurrence of large events.

The scorecard is one component of NASA's QuakeSim project. "QuakeSim seeks to develop tools for quake forecasting. It integrates high-precision, space-based measurements from Global Positioning System satellites and interferometric synthetic aperture radar with numerical simulations and pattern-recognition techniques," said JPL's Dr. Andrea Donnellan, QuakeSim principal investigator. "It includes historical data, geological information, and satellite data to make updated forecasts of quakes, similar to a weather forecast."

JPL software engineer Jay Parker said, "QuakeSim aims to accelerate the efforts of the international earthquake science community to better understand earthquake sources and develop innovative forecasting methods. We expect adding more types of data and analyses will

Image courtesy of the U.S. Geological Survey, C.E. Meyer.

Collapsed and burned buildings shown at Beach and Divisadero streets in the Marina District, San Francisco.

lead to forecasts with substantially better precision than we have today."

The scorecard forecast generated a map of California from the San Francisco Bay area to the Mexican border and then divided it into approximately 4,000 boxes, or "tiles." For each tile, researchers calculated the seismic potential and assigned color-coding to show the areas most likely to experience quakes over a 10-year period.

"In California, quake activity happens at some level almost everywhere. This method narrows the locations of the largest future events to about 6 percent of the state," Rundle said. "This information will help engineers and government decision makers prioritize areas for further testing and seismic retrofits."

So far, the technique has only missed one earthquake, a magnitude of 5.2, on June 15, 2004, under the ocean near San Clemente Island. Rundle believes this "miss" may be due to larger uncertainties in locating earthquakes in this offshore region of the state. San Clemente Island is at the edge of the coverage area for southern California's seismograph network. The research team is working to refine the method and find new ways to visualize the data.

Laser Technology Tracks Changes in Mount St. Helens

The U.S. Geological Survey (USGS) and NASA scientists studying Mount St. Helens are using high-tech light detection and ranging (lidar) technology to analyze changes in the surface elevation of the volcanic crater, which began deforming in late September 2004.

With data derived from airborne lidar, scientists can accurately map, often in exquisite detail, the dimensions of the uplift and create better models to forecast volcanic hazards. lidar shows, in the 2 weeks before October 4, the new uplift grew to the height of a 35-story building and the area of 29 football fields.

"This is the first time USGS and NASA have teamed to use lidar to measure volcano deformation," said USGS scientist Ralph Haugerud. He noted that lidar technology enables researchers to compare with greater accuracy than ever before the topography before and after volcanic events.

Mount St. Helens as seen from above. NASA is working with the U.S. Geological Survey to study the still-active volcano.

"The resulting pictures of topographic change can reveal information found in no other kind of data set," added David Harding, a scientist at Goddard.

In 2003, the USGS contracted a lidar survey of Mount St. Helens. In early September 2004, USGS and NASA scientists began detailed planning for a second survey. The survey, contracted by NASA, extended the area covered by the first survey. But when the mountain began rumbling on September 23, USGS and NASA scientists accelerated plans and re-surveyed the mountain on October 4.

Some of the Mount St. Helens features related to the volcanic unrest visualized in the new lidar-derived digital elevation model include growth of a new volcanic dome south of the 1980-1986 volcanic dome and new steam-and-ash vents.

Additional changes between the two lidar surveys unrelated to the volcanic unrest include shrinking snow fields, several rock falls, movement of three rock glaciers, and growth of the crater glacier, which has been an ongoing subject of USGS research at Mount St. Helens.

Lidar mapping uses a scanning laser rangefinder mounted in a small aircraft to measure distances from the aircraft to the ground several tens of thousands of times each second. It commonly measures the ground position at points a meter apart with vertical accuracy as good as 10 centimeters.

"NASA scientists and engineers in the 1980s and 1990s pioneered airborne lidar mapping," Harding said. "Because of its very high accuracy and fast turnaround of results, lidar is rapidly becoming the preferred method for detailed topographic mapping and is conducted worldwide on a commercial basis by numerous companies," he added.

Creating Earth's Most Extensive Global Topographic Map

Culminating more than 4 years of processing data, NASA and the National Geospatial-Intelligence Agency this year completed Earth's most extensive global topographic map.

The data, extensive enough to fill the U.S. Library of Congress, were gathered during Endeavour's Shuttle Radar Topography Mission, in 2000.

The digital elevation maps encompass 80 percent of Earth's landmass. They reveal for the first time large, detailed swaths of Earth's topography, previously obscured by persistent cloudiness. The data will benefit scientists, engineers, government agencies, and the public with an ever-growing array of uses.

"This is among the most significant science missions the Shuttle has ever performed, and it's probably the most significant mapping mission of any single type ever," said Dr. Michael Kobrick, the mission project scientist from JPL.

The final data release covers Australia and New Zealand in unprecedented uniform detail. It also covers more than 1,000 islands comprising much of Polynesia and Melanesia in the South Pacific, as well as islands in the South Indian and Atlantic oceans.

"Many of these islands have never had their topography mapped," Kobrick said.

The resulting data are being used for applications ranging from land-use planning to "virtual" Earth exploration. "Future missions using similar technology could monitor changes in Earth's topography over time, and even map the topography of other planets," said Dr. John LaBrecque, manager of NASA's Solid Earth and Natural Hazards Program, at NASA Headquarters.

Monitoring the Environment in Central America

A state-of-the-art environmental monitoring facility in Panama is the first to employ NASA Earth science research and space-based observations to provide Central American decision makers with early warning about a variety of ecological and climatic changes.

Developed by NASA scientists, the innovative regional monitoring system is called SERVIR, the Spanish acronym for the Regional Visualization and Monitoring System. It is also a Spanish term meaning "to serve."

Featuring a massive, Web-based data archive of maps and satellite imagery, decision-support tools, and interactive visualization capabilities, SERVIR is designed to aid government and industry across the seven countries of Central America and the southern Mexican states. "NASA's science mission begins here on Earth, with greater awareness and understanding of our changing planet, and new solutions for protecting our environment, resources, and human lives," said Dr. Ghassem Asrar, NASA's Deputy Associate Administrator for Science. "SERVIR technology, our partnership with various organizations, and with the people of Central America reflects NASA's commitment to improving life on our home planet for all people," he said.

The system contains user-friendly, interactive tools. It is designed to make NASA Earth observations and predictions freely and readily accessible to anyone with an Internet connection. Designed to track weather, climate, and ecological events, the system has already shown results in Central America, monitoring wildfires, red tides, and blooms of toxic algae threatening local

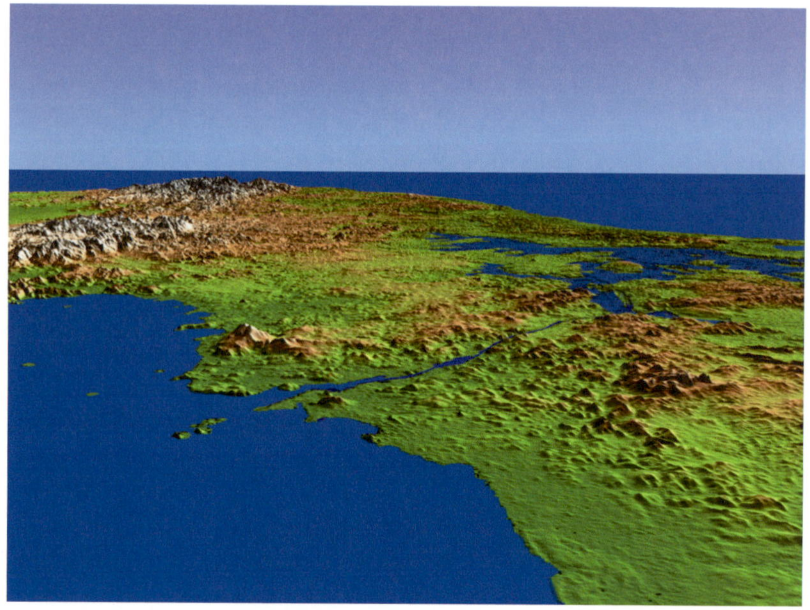

The SERVIR system is designed to make NASA Earth observations and predictions freely and readily accessible to anyone with an Internet connection. Designed to track weather, climate, and ecological events, the system has already shown results in Central America, monitoring wildfires, red tides, and blooms of toxic algae threatening local fishing areas.

fishing areas. "SERVIR is an excellent tool for gauging slow or periodic shifts in climate that could lead to drought and other long-term problems, as well as identifying quick-forming weather phenomena that threaten human lives and operations on land and at sea," said Daniel Irwin, SERVIR project manager at NASA's Marshall Space Flight Center.

NASA devised the system in partnership with the U.S. Agency for International Development; the World Bank; the City of Knowledge, Panama; the Water Center for the Humid Tropics of Latin America and the Caribbean (CATHALAC); the Central American Commission on Environment and Development; and Cable & Wireless Panama.

"We're extremely proud of this combined effort," said Tom Sever, SERVIR principal investigator at Marshall. "Without the partnership of these organizations, we never could have integrated the resources to create such a robust system—combining space-based observations with local knowledge of ecosystems to enable constant, real-time monitoring of this environmentally vital region."

The Panamanian SERVIR center is housed in the City of Knowledge, at CATHALAC. The City of Knowledge is an international consortium of health, science, and academic organizations including the United Nations Educational, Scientific and Cultural Organization; the World Food Programme; numerous universities; and research institutes.

NASA's Supercomputer Is Ranked Among the World's Fastest

NASA announced that its newest supercomputer, "Columbia," was named one of the world's most powerful production supercomputers by the TOP500 project at SC2004, the international conference of high-performance computing, networking, and storage.

Columbia, which achieved a benchmark rating of 51.9 teraflop/s on 10,240 processors, was ranked second on the TOP500 list, just behind "Blue Gene," IBM's supercomputer at the Department of Energy's Lawrence Livermore National Laboratory.

"Large, integrated simulation environments like those we have at Ames are crucial to NASA's missions, and Columbia has provided a breakthrough increase in our computational power," said G. Scott Hubbard, director of Ames. "A high rating on the TOP500 list is an impressive achievement, but for NASA, the immediate availability to analyze important issues like 'Return to Flight' for the Space Shuttle, space science, Earth modeling, and aerospace vehicle design for exploration, is the true measure of success."

"Columbia allows NASA to perform numerical simulations at the cutting edge of science and engineering," said Dr. Walter Brooks, chief of the NASA Advanced Supercomputing (NAS) Division at Ames. "As the largest example of an important high-end computing architecture developed in the U.S., part of this system will be available to the nation's best research teams. The swift design and deployment of Columbia has redefined the concept of supercomputer development."

With Columbia at its core, said Brooks, the NAS facility provides an integrated computing, visualization, and data storage environment to help NASA meet its mission goals and the Vision for Space Exploration.

Within days of completion of the supercomputer's installation, Columbia achieved a Linpack benchmark rating of 42.7 teraflop/s on just 16 nodes with an 88-percent efficiency rating, exceeding the previously best reported performance by a significant margin. This was followed almost immediately by a 51.7 teraflop/s rating reported for the entire system.

"What is most noteworthy is that we were able to post such a significant and efficient Linpack result in such a short time," said Bob Ciotti, chief systems engineer for the Columbia installation project. "Not only was the system deployed in less than 120 days, but the code used to achieve this result was conceived and developed in that same time frame, and is much more straightforward than the traditional approach. Our simplified implementation, allowed by shared memory systems like the SGI Altix, translates directly into improved effectiveness for users of our systems."

The supercomputer was named to honor the crew of the Space Shuttle Columbia, lost February 1, 2003.

Debugging Computer Code

NASA scientists announced in April that they are releasing free software that will find "bugs," or defects, in Java computer code.

The new software, Java PathFinder, is classified as open source software, which is computer code that scientists make publicly available, often at no cost, so users can freely utilize and modify it. Java is a computer language that software developers frequently use to write programs for computer networks, such as the Internet.

According to John Penix, a computer scientist at Ames, the open source offering will enable other people to help Ames improve the PathFinder software for NASA's benefits, too.

The Java PathFinder work "is part of an effort to develop tools and methods to identify and eliminate software errors in NASA's increasingly complex and

Ames Research Center's Willem Visser stands in front of a projection of a Java PathFinder software computer screen.

mission-critical software systems," said David Korsmeyer, who leads Ames's Intelligent Systems Division. "Java PathFinder was used to detect inconsistencies in the executive software for the K9 Rover at NASA Ames," Korsmeyer added. The K9 is a six-wheeled, solar-powered rover developed jointly at Ames and JPL.

In addition, computer scientists used elements of Java PathFinder to develop verification computer code for Livingstone 2 software, a diagnosis system that is now flying on the EO-1 spacecraft and, "an example of the kind of autonomy software that will be crucial to future NASA missions," emphasized Korsmeyer.

According to scientists, if PathFinder finds an error in a Java application, the software checker reports the whole process that leads to the bug. "Unlike a normal debugger, Java PathFinder keeps track of every step the software checker takes to find a defect," Penix noted.

"PathFinder already has been enhanced and tested by several universities and companies," Penix said. "Now, additional universities can add more features to PathFinder."

The software is in its sixth year of active development.

High-Tech X-Ray Equipment Examines Dinosaur Skull

As NASA charts a bold new course into the future, the Space Agency is briefly taking a step back in time to examine a dinosaur skull. NASA scientists are using equipment at Marshall to scan the skull of a Tyrannosaurus rex (*T. rex*). The state-of-the-art equipment was originally designed to examine rocket motor assemblies and turbine blades. Discovered on a South Dakota ranch in 1992, it is believed to be the most complete and well-preserved *T. rex* skull ever found. Discoverers dubbed the find "Samson," recognizing the beast's reputation as the strongest dinosaur to roam the Earth during the late Cretaceous period.

"Marshall is one of the few places in the world with the technology needed for such a complex scan," said Dr. Chris Beard, curator of vertebrate paleontology at the Carnegie Museum of Natural History in Pittsburgh. "We are very excited NASA has agreed to provide space-age technology for this project."

Dr. Ron Beshears is leading the project at the National Center for Advanced Manufacturing located at Marshall. Beshears's laboratory team is running various tests on the skull with a high-tech computed tomography scanner used for nondestructive testing of parts and equipment destined for space. The scans provide Carnegie Museum experts with detailed cross-section images of the skull.

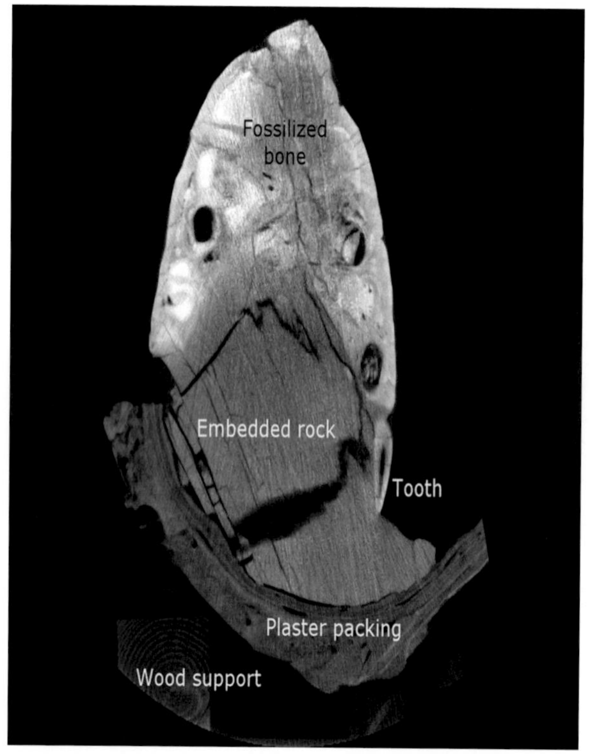

A fossilized Tyrannosaurus rex skull is being tested by researchers at Marshall, using a high-tech tomography scanner originally designed to examine rocket motor assemblies and turbine blades.

Such detail will help museum experts better understand the basic anatomy and lifestyle of the *T. rex*.

"The idea of working with 65 million-year-old dinosaur bones alongside next-generation space technologies is something we're quite excited about," Beshears said. "We're happy we can use our facility to assist in a scientific investigation of the dinosaur fossil."

Carnegie Museum researchers will use results to compare Samson's skull with previous computed tomography scans of less well-preserved *T. rex* fossils, establishing a baseline to determine anomalies in future finds. Although privately owned, Samson is being prepared and studied by the museum for 2 years. The dinosaur arrived at the Carnegie Museum in May 2004.

The skull, separated from its skeleton by the museum for study, is largely encrusted in rock. It arrived at Marshall enclosed in a shipping crate approximately 5 feet by 3.5 feet and weighed approximately 1,600 pounds. Because of the skull's fragility, it will not be removed from the crate while tests are performed. After tests and examinations are completed, it will be returned to the Carnegie Museum to recreate the once-fearsome predator.

Eradicating Invasive Species

In February, NASA accepted an invitation to join the National Invasive Species Council (NISC), to assist 12 other Federal agencies in combating invasive species across the country.

An invasive species is an organism, such as a microbe, plant, or animal, which entered America through natural processes or with human assistance and whose presence poses a threat to public health or the economy. One example, salt cedar, is an invasive plant widespread in the western United States. It replaced native species and may have significant negative effects on water resources.

"NASA is pleased with this invitation from NISC. The agency is eager to continue our active engagement in applied research projects whose results advance management of invasive species," said Edwin Sheffner, manager

Image courtesy of David Cappaert, www.forestryimages.org and www.invasive.org.

The Emerald Ash Borer is an exotic beetle that was discovered in southeastern Michigan near Detroit in the summer of 2002. Since its discovery, the invasive species has killed at least 8 to 10 million ash trees in Michigan, Ohio, and Indiana.

of the invasive species program element in the Applied Sciences Program at NASA Headquarters. "Efforts to manage invasive species annually cost the country tens of billions of dollars," he said.

NASA will enhance its partners' abilities to respond effectively and efficiently to invasive species' challenges. Its track record of achievement in invasive species monitoring led to the invitation to join the council.

An example of the Space Agency's impact is work completed by the USGS on invasive species in Utah's Grand Staircase-Escalante National Monument. The USGS improved the accuracy and timeliness of predictive maps of invasive species in the monument with enhancements to decision support tools from NASA data, predictive models, and systems engineering.

MODIS data from NASA's Terra satellite provides daily information about vegetation conditions. Statistical models applied by the USGS, with NASA's assistance, convert MODIS and other data sources into predictive maps of plant species distribution. The USGS is incorporating NASA's research capabilities to improve the national response to invasive species through the National Invasive Species Forecasting System.

NASA Technology Supports Virgin Atlantic GlobalFlyer

GlobalFlyer landed safely in Salina, Kansas, after the first solo, non-stop, non-refueled, around-the-world airplane trip. NASA technology contributed to the safety and success of the mission by enhancing communications between pilot Steve Fossett and his ground control team. NASA's real-time video hookup allowed aviation enthusiasts around the globe to follow the landmark flight.

The flight tested NASA's advanced experimental Tracking and Data Relay Satellite System (TDRSS) transceiver, called the Low Power Transceiver (LPT). As a side benefit, the NASA device allowed GlobalFlyer's mission control to communicate with Fossett for almost 3 days of flight through a live video connection.

"We at NASA applaud private sector record-setting achievements like this one. NASA is committed to increasing its engagement with entrepreneurs and industry alike in pursuit of the Vision for Space Exploration," said NASA's Associate Administrator for Space Operations, William Readdy. "We're proud of our very talented, dedicated people and cutting-edge technologies and look forward to even more partnering in the future."

NASA researchers believe the LPT holds promise as a flexible and less expensive option for relaying information to and from spacecraft. TDRSS already supports space operations by providing uninterrupted data relay and communications between orbiting spacecraft and the ground.

From the LPT testing on GlobalFlyer, NASA hopes to learn more about how the device operated during flight, especially when transmitting video. Four NASA facilities, Goddard Space Flight Center, Dryden Flight Research Center, Kennedy Space Center, and the White Sands Test Facility, contributed technology to the video project and monitored the GlobalFlyer mission.

NASA also loaned GlobalFlyer its Personal Cabin Pressure Monitor, which alerts a pilot of potentially dangerous or deteriorating cabin pressure. Because Fossett's cockpit was exceedingly loud, too loud for an alarm, the device was modified to vibrate when signaling a problem.

Preventing Air Traffic Bottlenecks

No one is happy with long lines and delays at our Nation's airports. In response to the growing need to improve the National Airspace System, NASA is developing tools to ensure future air travel will be safe and efficient.

NASA, the Federal Aviation Administration (FAA), and the MITRE Corporation, of McLean, Virginia, successfully conducted tests of the Multi-Center Traffic Management Advisor (McTMA) at air traffic facilities responsible for the northeastern United States. Initial results indicate the software's scheduling capabilities helped air traffic managers prevent bottlenecks.

At the heart of McTMA is a powerful "trajectory synthesis" engine capable of converting radar data, flight plans, and weather information into accurate forecasts of air traffic congestion. McTMA uses these forecasts and input from air traffic personnel to generate a specific advisory, typically a small delay, for each aircraft predicted to encounter congestion.

"McTMA is an advanced air traffic management system that makes possible a fundamental shift in air traffic control form from distance-based to time-based metering of aircraft," said Tom Edwards, deputy director of the Aeronautics Directorate at Ames. "Time-based metering can reduce airborne delays and improve coordination and planning between adjacent air traffic control facilities," he added.

Tests were conducted with managers at the Air Route Traffic Control Centers in New York, Washington, Boston, and Cleveland, as well as the Philadelphia Terminal Radar Approach Control and the National Air Traffic Control System Command Center in Herndon, Virginia.

The successful tests validated the McTMA "distributed scheduling architecture," and helped air traffic managers prevent bottlenecks at the Philadelphia International Airport.

NASA researchers are hard at work to make airline travel safer and more efficient.

"The evaluation successfully demonstrated the advantages of the McTMA departure metering capability over current techniques," said Tom Davis, principal investigator for McTMA and chief of the Terminal Area Air Traffic Management Research Branch at Ames. "During several periods at Philadelphia, when airborne holding is routinely encountered, no such holding was observed when McTMA was in use," Davis added.

Frequently, adjustments of just a few minutes at the point of origin can alleviate airborne traffic jams at the destination. The result is safer and more efficient operations for airlines and the flying public as the system produces a steady but manageable flow of air traffic.

"Future tests will seek to gradually expand the McTMA operational envelope to demonstrate multi-center, time-based metering of departures, arrivals and en route flows to multiple destinations," Davis said.

Earlier versions of the system are used to schedule arriving aircraft at Dallas-Ft. Worth, Minneapolis, Los Angeles, Denver, Houston, Miami, and Atlanta airports.

Testing of the newer McTMA system continues in 2005 at the same facilities. If fully successful, NASA and the FAA will work together to bring the technology into future operations to benefit air travelers.

The program is managed by the Airspace Systems Division of NASA's Aeronautics Research Mission Directorate. The software was developed at Ames.

NASA Deploys ChemSecure Hazmat Management System

Dryden Flight Research Center is implementing an extensive wireless, sensor-based system aimed at improving the management of hazardous materials to enhance security and safety, while significantly reducing ongoing supply chain costs. The ChemSecure pilot program integrates radio frequency identification (RFID) and sensor-based technology with the U.S. Department of Defense's existing Web-based Hazardous Materials Management System (HMMS) database to automate the real-time management of hazardous materials including usage, shipment, tracking, and storage.

The first project of its kind, Dryden developed ChemSecure in close partnership with the Department of Defense and leading private sector companies.

"The ChemSecure program is a testament to NASA's commitment to using advanced technology and business processes to create safer, more secure management systems for hazardous material movement and storage," said Ralph Anton, chemical program manager at Dryden.

"ChemSecure's guiding business processes and technology foundation are not limited to the hazardous materials environment—we see numerous applications for tracking of a variety of materials, in the public and private sectors, and plan to help agencies and organizations take advantage of the system."

ChemSecure places RFID tags on hazardous material containers and uses Oracle Sensor-Based Services to capture, manage, analyze, and respond to any movement or other change of the chemicals. Dryden applies the real-time information in the HMMS database to make informed decisions about the transportation and storage of hazardous materials, and provides automatic alerts—text messaging, voice alerts, and e-mails—to professionals in security, safety, health, and environment to warn them of any changes with the chemicals.

In addition to helping organizations significantly reduce hazardous materials management costs and errors, the ChemSecure program includes many additional capabilities that enhance safety and security measures such as:

- Supplying critical data to first responders and decision makers so they are equipped to make timely decisions for the safety, security, and protection of people, as well as the physical assets in the environment during an emergency evacuation involving a chemical spill;
- Monitoring personnel when they handle hazardous containers and providing accountability by crosschecking personnel information with container information to reduce theft, error, and fraud;
- Providing end-to-end visibility of the hazardous materials transportation and storage life cycle for improved decision making and auditing;
- Ensuring chemicals are placed in appropriate and safe locations to avoid adverse reactions with other chemicals; and
- Making sure personnel are properly authorized and trained to work with the chemicals to reduce human error.

Dryden is planning a second phase of the ChemSecure project that will provide enhanced features for scrutinizing all vehicles entering and leaving unguarded access points; and for maintaining full inventory management throughout the facility, extending the homeland security element of this project. Additionally, the sensor-based technology will track all climate-controlled chemicals in restricted environments. ❖

NASA Dryden chemical crib technician Christina Urias first enters information about the hazardous material into the Hazardous Materials Management System, allowing the radio frequency identification sensor tag, affixed to each container, to send accurate information to emergency responders.

TECHNOLOGY TRANSFER NETWORK

AND AFFILIATIONS

The Innovative Partnerships Program creates alliances with industry, academia, and other non-traditional sources to develop and transfer technology in support of national priorities and NASA's missions. The programs and activities resulting from the alliances engage innovators and enterprises to fulfill NASA's mission needs and promote the potential of NASA technology. The following section recognizes all of the organizations that assist in establishing these fruitful relationships.

2005 Technology Transfer Network and Affiliations

The NASA Innovative Partnerships Program sponsors a number of organizations around the country that are designed to assist U.S. businesses in accessing, utilizing, and commercializing NASA-funded research and technology. These organizations work closely with the Technology Transfer Offices, located at each of the 10 NASA field centers, to provide a full range of technology transfer and commercialization services and assistance.

Technology Transfer Network

The **National Technology Transfer Center (NTTC) <http://www.nttc.edu>**, located on the campus of Wheeling Jesuit University, in Wheeling, West Virginia, was established by the U.S. Congress in 1989 to strengthen American industry by providing access to more than $70 billion worth of federally funded research. By helping American companies use Federal technologies, the NTTC helps them manufacture products, create jobs, and foster partnerships between Federal laboratories and the private sector, universities, innovators, and economic development organizations. From that mission, the NTTC has grown into a full-service technology commercialization center. In addition to providing access to Federal technology information, the NTTC provides technology commercialization training; technology assessment services that help guide industries in making key decisions regarding intellectual property and licensing; and assistance in finding strategic business partners and electronic business development services.

The NTTC developed a leads management system for NASA that is the formal reporting and tracking system for partnerships being developed between NASA and U.S. industry. The leads system allows all members of the NASA Technology Commercialization Team to have an easy-to-use and effective tool to create and track leads in order to bring them to partnerships. The NTTC also utilizes the expertise of nationally recognized technology management experts to create and offer technology commercialization training. Course topics range from the basics of technology transfer to hands-on valuation, negotiation, and licensing. Courses are developed at the NTTC and around the country. In addition, online courses, supporting publications, comprehensive software applications, and videotapes are also available.

The National Technology Transfer Center is located on the campus of Wheeling Jesuit University, in Wheeling, West Virginia.

NASA TechTracS <http://technology.nasa.gov> provides access to NASA's technology inventory and numerous examples of the successful transfer of NASA-sponsored technology for commercialization. TechFinder, the main feature of the Internet site, allows users to search technologies and success stories, as well as submit requests for additional information. All NASA field centers submit information to the TechTracS database as a means of tracking technologies that have potential for commercial development.

Since their inception in January 1992, the six NASA-sponsored Regional Technology Transfer Centers (RTTCs) have helped U.S. businesses investigate and utilize NASA and other federally funded technologies for companies seeking new products, improvements to existing products, or solutions to technical problems. The RTTCs provide technical and business assistance to several thousand customers every year.

The network of RTTCs is divided as follows: **Far West** (AK, AZ, CA, HI, ID, NV, OR, WA): The **Far West Regional Technology Transfer Center (FWRTTC) <http://www.usc.edu/dept/engineering/TTC/NASA>** is an engineering research center within the School of Engineering at the University of Southern California in Los Angeles. Using the Remote Information Service to generate information from hundreds of Federal databases, FWRTTC staff work closely with businesses and entrepreneurs to identify opportunities, expertise, and other necessary resources. The FWRTTC enhances the relationships between NASA and the private sector by offering many unique services, such as the NASA Online Resource Workshop, NASA Tech Opps, and links to funding and conference updates.

Mid-Atlantic (DC, DE, MD, PA, VA, WV): The **Technology Commercialization Center (TeCC) <http://www.teccenter.org>**, located in Hampton, Virginia, coordinates and assists in the transfer of marketable technologies, primarily from Langley Research Center, to private industry interested in developing and commercializing new products.

Mid-Continent (AR, CO, IA, KS, MO, MT, ND, NE, NM, OK, SD, TX, UT, WY): The **Mid-Continent Technology Transfer Center (MCTTC) <http://www.mcttc.com/>**, under the direction of the Technology and Economic Development Division of the Texas Engineering Service, is located in College Station, Texas.

The MCTTC, which provides a link between private companies and Federal laboratories, reports directly to the Johnson Space Center. The assistance focuses on high-tech and manufacturing companies that need to acquire and commercialize new technology.

Mid-West (IL, IN, MI, MN, OH, WI): The **Great Lakes Industrial Technology Center (GLITeC) <http://www.glitec.org>**, managed by Battelle Memorial Institute, is located in Cleveland, Ohio. GLITeC works with industries primarily within its six-state region to acquire and use NASA technology and expertise through the partnership with the Glenn Research Center. Each year, over 500 companies work with GLITeC and its affiliates to identify new market and product opportunities. Technology-based problem solving, product planning and development, and technology commercialization assistance are among the services provided.

Northeast (CT, MA, ME, NH, NJ, NY, RI, VT): The **Center for Technology Commercialization (CTC) <http://www.ctc.org>** is a nonprofit organization, based in Westborough, Massachusetts. Covering New England, New York, and New Jersey, the CTC currently has seven satellite offices that form strong relationships with Northeast industry. Operated by the CTC, the NASA Business Outreach Office stimulates business among regional contractors, NASA field centers, and NASA prime contractors.

Southeast (AL, FL, GA, KY, LA, MS, NC, SC, TN): The **Southeast Regional Technology Transfer Center (SERTTC) <http://www.edi.gatech.edu/nasa>** at the Georgia Institute of Technology facilitates and coordinates private industry interests in the transfer and commercialization of technologies resulting from NASA's space and Earth science research. Assistance is also provided in Small Business Innovation Research and Small Business Technology Transfer applications, as well as the establishment of connections to specialized research needs within NASA research and development centers nationwide.

NASA Incubator Programs

Eight NASA incubators are included within this network of programs. They are designed to nurture new and emerging businesses with the potential to incorporate technology developed by NASA. They offer a wide variety of business and technical support services to increase the success of participating companies.

BizTech <http://www.biztech.org>, of Huntsville, Alabama, is a small business incubator, offering participating companies access to services at Marshall Space Flight Center laboratories for feasibility testing, prototype fabrication, and advice on technology usage and transfer. BizTech is sponsored by the Huntsville-Madison County Chamber of Commerce.

The **Emerging Technology Centers (ETC) <http://www.etcbaltimore.com>**, located in Baltimore, Maryland, is one of the newest NASA-affiliated incubators. Partnering institutions include the Goddard Space Flight Center and area universities and colleges.

The **Florida/NASA Business Incubation Center (FNBIC) <http://www.trda.org/fnbic/>** is a joint partnership of NASA's Kennedy Space Center, Brevard Community College, and the Technological Research and Development Authority. The mission of the FNBIC is to increase the number of successful technology-based small businesses originating in, developing in, or relocating to Brevard County. The FNBIC offers support facilities and programs to train and nurture new entrepreneurs in the establishment and operation of developing ventures based on NASA technology.

The **Hampton Roads Technology Incubator (HRTI) <http://www.hr-incubator.org>** identifies and licenses Langley Research Center technologies for commercial use. The HRTI's mission is to increase the number of successful technology-based companies originating in, developing in, or relocating to the Hampton Roads area.

The **Lewis Incubator for Technology (LIFT) <http://www.liftinc.org>**, managed by Enterprise Development, Inc., provides outstanding resources for technology and support to businesses in the Ohio region. Its primary objectives are to create businesses and jobs in Ohio and to increase the commercial value of NASA knowledge, technology, and expertise. LIFT offers a wide range of services and facilities to the entrepreneur to increase the probability of business success.

The **Mississippi Enterprise for Technology (MsET) <http://www.mset.org>** is sponsored by NASA and the Mississippi University Consortium and Department of Economic and Community Development, as well as the private sector. The mission of the enterprise is to help small businesses utilize the scientific knowledge and technical expertise at the Stennis Space Center. A significant part of this effort is Stennis' Commercial Remote Sensing program, which was formed to commercialize remote sensing, geographic information systems, and related imaging technologies.

The **NASA Commercialization Center (NCC) <http://www.nasaincubator.csupomona.edu>**, run by California State Polytechnic University, Pomona, is a business incubator dedicated to helping small businesses access and commercialize Jet Propulsion Laboratory and Dryden Flight Research Center technologies.

The **UH-NASA Technology Commercialization Incubator <http://www.research.uh.edu>** is a partnership between Johnson Space Center and the University of Houston. The incubator is designed to help local small and mid-size Texas businesses commercialize space technology. The University of Houston houses the program and provides the commercialization and research expertise of its business and engineering faculties.

Affiliated Organizations, Services, and Products

To complement the specialized centers and programs sponsored by the NASA Innovative Partnerships Program, affiliated organizations and services have been formed to strengthen NASA's commitment to U.S.

businesses. Private and public sector enterprises build upon NASA's experience in technology transfer in order to help with the channeling of NASA technology into the commercial marketplace.

The NASA **Small Business Innovation Research (SBIR)** program **<http://www.sbir.nasa.gov>** provides seed money to U.S. small businesses for developing innovative concepts that meet NASA mission requirements. Each year, NASA invites small businesses to offer proposals in response to technical topics listed in the annual SBIR program solicitation. The NASA field centers negotiate and award the contracts, as well as monitor the work.

NASA's SBIR program is implemented in three phases:

- **Phase I** is the opportunity to establish the feasibility and technical merit of a proposed innovation. Selected competitively, NASA Phase I contracts last 6 months and must remain under specific monetary limits.
- **Phase II** is the major research and development effort which continues the most promising of the Phase I projects based on scientific and technical merit, results of Phase I, expected value to NASA, company capability, and commercial potential. Phase II places greater emphasis on the commercial value of the innovation. The contracts are usually in effect for a period of 24 months and again must not exceed specified monetary limits.
- **Phase III** is the process of completing the development of a product to make it commercially available. While the financial resources needed must be obtained from sources other than the funding set aside for the SBIR, NASA may fund Phase III activities for follow-on development or for production of an innovation for its own use.

The SBIR Management Office, located at the Goddard Space Flight Center, provides overall management and direction of the SBIR program.

The NASA **Small Business Technology Transfer (STTR)** program **<http://www.sbir.nasa.gov>** awards contracts to small businesses for cooperative research and development with a research institution through a uniform, three-phase process. The goal of Congress in establishing the STTR program was to transfer technology developed by universities and Federal laboratories to the marketplace through the entrepreneurship of a small business.

Although modeled after the SBIR program, STTR is a separate activity and is separately funded. The STTR program differs from the SBIR program in that the funding and technical scope is limited and participants must be teams of small businesses and research institutions that will conduct joint research.

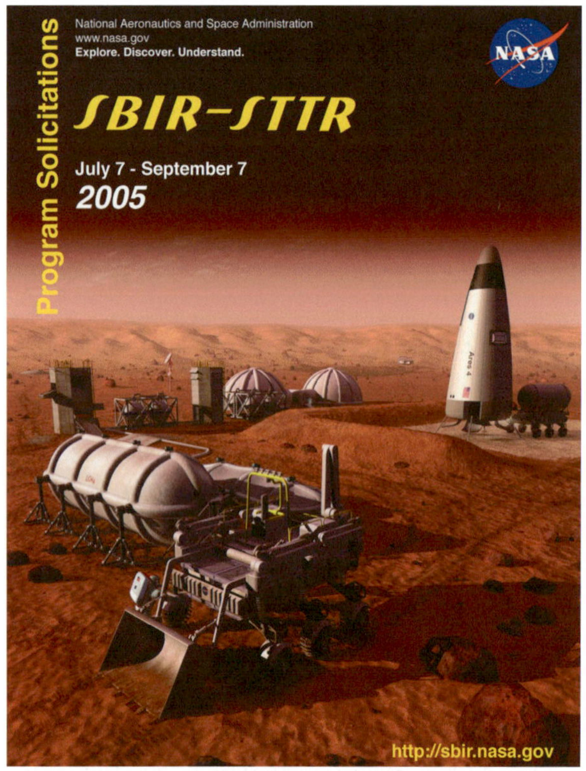

NASA's Small Business Innovation Research (SBIR) and Small Business Technology Transfer (STTR) programs are a three-phased approach for the small business concern to develop a technology in response to a specific set of NASA mission-driven needs, as presented in the NASA SBIR/STTR Annual Solicitation.

The **Federal Laboratory Consortium (FLC) for Technology Transfer <http://www.federallabs.org>** was organized in 1974 to promote and strengthen technology transfer nationwide. More than 600 major Federal laboratories and centers, including NASA, are currently members. The mission of the FLC is twofold:

- To promote and facilitate the rapid movement of Federal laboratory research results and technologies into the mainstream of the U.S. economy.
- To use a coordinated program that meets the technology transfer support needs of FLC member laboratories, agencies, and their potential partners in the transfer process.

The **National Robotics Engineering Consortium (NREC) <http://www.rec.ri.cmu.edu>** is a cooperative venture among NASA, the City of Pittsburgh, the State of Pennsylvania, and Carnegie Mellon's Robotics Institute. Its mission is to move NASA-funded robotics technology to industry. Industrial partners join the NREC with the goal of using technology to gain a greater market share, develop new niche markets, or create entirely new markets within their area of expertise.

Other organizations devoted to the transfer of NASA technology are the **Research Triangle Institute (RTI)** and the **MSU TechLink Center**.

The **RTI <http://www.rti.org>**, located in Research Triangle Park, North Carolina, provides a range of technology management services to NASA. RTI performs technology assessments to determine applications and commercial potential of NASA technology, as well as market analysis, and commercialization and partnership development. RTI works closely with all of NASA's Technology Transfer Offices.

The National Robotics Engineering Consortium, located in Pittsburgh, develops and applies advanced automation technologies to create new opportunities for American businesses.

The **MSU TechLink Center <http://techlink.msu.montana.edu>** located at Montana State University-Bozeman, was established in 1997 to match the technology needs of client companies with resources throughout NASA and the Federal laboratory system. TechLink focuses on a five-state region that includes Idaho, Montana, North Dakota, South Dakota, and Wyoming. Working closely with public, private, and university programs, TechLink provides ongoing support in the process of adapting, integrating, and commercializing NASA technology.

The road to technology commercialization begins with the basic and applied research results from the work of scientists, engineers, and other technical and management personnel. The NASA **Scientific and Technical Information (STI)** program **<http://www.sti.nasa.gov>** provides the widest appropriate dissemination of NASA's research results. The STI program acquires, processes, archives, announces, and disseminates NASA's internal—as well as worldwide—STI.

The NASA STI program offers users Internet access to its database of over 3.9 million citations, as well as many in full text; online ordering of documents; and the NASA STI Help Desk (help@sti.nasa.gov) for assistance in accessing STI resources and information. Free registration with the program is available through the NASA Center for AeroSpace Information.

For more than 3 decades, reporting to industry on any new, commercially significant technologies developed in the course of NASA research and development efforts has been accomplished through the publication of ***NASA Tech Briefs*** **<http://www.nasatech.com>**.

The monthly magazine features innovations from NASA, industry partners, and contractors that can be applied to develop new or improved products and solve engineering or manufacturing problems. Authored by the engineers or scientists who performed the original work, the briefs cover a variety of disciplines, including computer software, mechanics, and life sciences. Most briefs offer a free supplemental technical support package, which explains the technology in greater detail and provides contact points for questions or licensing discussions.

Technology Innovation **<http://www.ipp.hq.nasa.gov/innovation/index.html>** is published quarterly by the NASA Office of Exploration Systems. Regular features include current news and opportunities in technology transfer and commercialization, and innovative research and development.

NASA *Spinoff* <http://www.sti.nasa.gov/tto/spinoff.html> is an annual print and online publication featuring successful commercial and industrial applications of NASA technology, current research and development efforts, and the latest developments from the NASA Innovative Partnerships Program. ❖

Technology Innovation is one of NASA's magazines for business and technology, published by the Innovative Partnerships Program. The latest issue highlights how the Vision for Space Exploration will bring great discoveries and benefits to all mankind.

FY 2005 NASA Technology Transfer Network

The FY 2005 NASA Technology Transfer Network (NTTN) extends from coast to coast. For specific information concerning commercial technology activities described below, contact the appropriate personnel at the facilities listed or go to the Internet at: **<http://www.ipp.nasa.gov>**. General inquiries may be forwarded to the National Technology Transfer Center at technology@nttc.edu.

To publish your success about a product or service you may have commercialized using NASA technology, assistance, or know-how, contact the NASA Center for AeroSpace Information or go to the Internet at: **<http://www.sti.nasa.gov/tto/contributor.html>**.

★ **NASA Headquarters** manages the Spinoff Program.

▲ **Field Center Technology Transfer Offices** represent NASA's technology sources and manage center participation in technology transfer activities.

✖ **National Technology Transfer Center (NTTC)** provides national information, referral, and commercialization services for NASA and other government laboratories.

● **Regional Technology Transfer Centers (RTTC)** provide rapid access to information, as well as technical and commercialization services.

■ **Research Triangle Institute (RTI)** provides a range of technology management services including technology assessment, valuation and marketing, market analysis, intellectual property audits, commercialization planning, and the development of partnerships.

★ NASA Headquarters

National Aeronautics and Space Administration
300 E Street, SW
Washington, DC 20546
NASA Spinoff Program Manager:
Janelle Turner
Phone: (202) 358-0704
E-mail: Janelle.B.Turner@nasa.gov

▲ Field Centers

Ames Research Center
National Aeronautics and Space Administration
Moffett Field, California 94035
Chief, Technology Partnerships Division:
Lisa Lockyer
Phone: (650) 604-1754
E-mail: lisa.l.lockyer@nasa.gov

Dryden Flight Research Center
National Aeronautics and Space Administration
4800 Lilly Drive, Building 4839
Edwards, California 93523-0273
Chief, Technology Transfer Partnerships Office:
Michael Gorn
Phone: (661) 276-2355
E-mail: michael.gorn@dfrc.nasa.gov

John H. Glenn Research Center at Lewis Field
National Aeronautics and Space Administration
21000 Brookpark Road
Cleveland, Ohio 44135
Director, Technology Transfer and Partnering Office:
Robert Lawrence
Phone: (216) 433-3484
E-mail: Robert.S.Lawrence@nasa.gov

Goddard Space Flight Center
National Aeronautics and Space Administration
Greenbelt, Maryland 20771
Chief, Office of Technology Transfer:
Nona K. Cheeks
Phone: (301) 286-5810
E-mail: Nona.K.Cheeks@nasa.gov

Jet Propulsion Laboratory
4800 Oak Grove Drive
Pasadena, California 91109
Manager, Commercial Program Office:
James K. Wolfenbarger
Phone: (818) 354-3821
E-mail: James.K.Wolfenbarger@nasa.gov

Lyndon B. Johnson Space Center
Houston, Texas 77058
Acting Director, Technology Transfer Office:
Helen Lane
Phone: (281) 483-7165
E-mail: helen.w.lane@nasa.gov

John F. Kennedy Space Center
National Aeronautics and Space Administration
Kennedy Space Center, Florida 32899
Chief, Technology Transfer Office:
James A. Aliberti
Phone: (321) 867-6224
E-mail: Jim.Aliberti@nasa.gov

Langley Research Center
National Aeronautics and Space Administration
Hampton, Virginia 23681-2199
Acting Deputy Head, Innovation Institute:
Richard R. Antcliff
Phone: (757) 864-3000
E-mail: Richard.R.Antcliff@nasa.gov

George C. Marshall Space Flight Center
National Aeronautics and Space Administration
Marshall Space Flight Center, Alabama 35812
Manager, Technology Transfer Office:
Vernotto C. McMillan
Phone: (256) 544-4266
E-mail: Vernotto.C.McMillan@nasa.gov

John C. Stennis Space Center
National Aeronautics and Space Administration
Stennis Space Center, Mississippi 39529
Chief, Technology Development and Transfer Office:
John W. Bailey
Phone: (228) 688-1065
E-mail: John.W.Bailey@nasa.gov

✖ National Technology Transfer Center (NTTC)

Wheeling Jesuit University
Wheeling, West Virginia 26003
Davitt McAteer, Vice President of Sponsored Programs
Phone: (800) 678-6882
E-mail: dmcateer@nttc.edu

● Regional Technology Transfer Centers (RTTCs)

Far West
Technology Transfer Center
3716 South Hope Street, Suite 200
Los Angeles, California 90007-4344
Kenneth Dozier, Director
Phone: (800) 642-2872
E-mail: kdozier@mizar.usc.edu

Mid-Atlantic
Technology Commercialization Center
144 Research Drive
Hampton, Virginia 23666
Duncan McIver, Director
Phone: (757) 766-9200
E-mail: dmciver@teccenter.org

Mid-Continent
Technology Transfer Center
Technology & Economic Development Division
Texas Engineering Extension Service
Connally Building
301 Tarrow MS 8000
College Station, Texas 77840-7896
Gary Sera, Director
Phone: (979) 845-8762
E-mail: gary.sera@teexmail.tamu.edu

Mid-West
Great Lakes Industrial Technology Center (GLITeC)
20445 Emerald Parkway Drive, SW, Suite 200
Cleveland, Ohio 44135
Marty Kress, President
Phone: (216) 898-6400
E-mail: kressm@batelle.org

Northeast
Center for Technology Commercialization (CTC)
1400 Computer Drive
Westborough, Massachusetts 01581
James P. Dunn, Director
Phone: (508) 870-0042
E-mail: jdunn@ctc.org

Southeast
Technology Transfer Center (SERTTC)
151 6th Street, 216 O'Keefe Building
Atlanta, Georgia 30332
David Bridges, Director
Phone: (404) 894-6786
E-mail: david.bridges@edi.gatech.edu

■ Research Triangle Institute (RTI)

Technology Application Team
3040 Cornwallis Road, P.O. Box 12194
Research Triangle Park, North Carolina 27709-2194
Dan Winfield, Director
Phone: (919) 541-6431
E-mail: winfield@rti.org

NASA Center for AeroSpace Information

Spinoff Project Office
NASA Center for AeroSpace Information
7121 Standard Drive
Hanover, Maryland 21076-1320

Jutta Schmidt, Project Manager
Phone: (301) 621-0182
E-mail: jschmidt@sti.nasa.gov

Jamie Janvier, Editor/Writer
Phone: (301) 621-0242
E-mail: jjanvier@sti.nasa.gov

Daniel Lockney, Editor/Writer
Phone: (301) 621-0244
E-mail: dlockney@sti.nasa.gov

John Jones, Graphic Designer

Deborah Drumheller, Publications Specialist